ISBN: 978-81-948672-7-2

Sustainable Agricultural Development in India: Exploring the Possibilities

Peer reviewed edited book

Edited by

Prof. Mahesha, M
Dr. Shashikumar, T.P.
Dr. Manjunatha, C.S

Published by Ambisphere Publications©
2022

Sustainable Agricultural Development in India: Exploring the Possibilities

Edited by
Prof. Mahesha, M
Dr. Shashikumar,T.P.
Dr. Manjunatha, C.S

Published by **Ambisphere Publications**
A Worldwide publishing and distribution house with multi-lingual publishing platforms.
Government of India recognized Publisher

Head Office: #13 old LIC colony, Jayanagar, Bengaluru (Bangalore) Urban, Karnataka, 560011, India.
Branches at: Mysore, Hyderabad, New Delhi, Mumbai
International Office and Collaborations: Dubai, Finland, Russia

PEER REVIEWED EDITED BOOK

This is peer reviewed edited book compiled and edited by the editors. The chapters in this book are written by the different authors. The content of the book is sole rights of editors and authors. All the information related to chapter should be contacted to individual authors. Publishers is not responsible for the content mentioned in the chapters.

Cover by Ambisphere Publications.

ISBN: 978-81-948672-7-2

ISBN: 978-81-948672-7-2

SUSTAINABLE AGRICULTURAL DEVELOPMENT IN INDIA:
EXPLORING THE POSSIBILITIES

TABLE OF CONTENTS

Foreward

Agriculture has been the main occupation of humanity since immemorial times. Indian agriculture has been practiced going back to 5000 B. C. As science and technology develop, the operation and cultivation methods change all over the world. Since the independence of the sovereign state of India, structural changes have taken place in all sectors. India, at the beginning of the planning era, adopted a five-year planning approach for its development. However, as developed countries implemented globalisation policies in their economic development, India followed suit. As economists and other leaders noticed, the over-production and over-consumption of natural resources as the world's population grew more rapidly. Even though man is an intelligent being, he does not realise the interdependency of nature and man on this earth.

In the name of economic growth and the wrong notion that man can do anything on this earth, he has started the exploitation of natural resources for his own advantage, which has resulted in unsustainability. World leaders and economists and scientific community felt that if man has continued to practice unsustain economic growth which has consequences on climate and environment, so that world leaders have adopted sustainable economic development strategy as the economic growth and development strategy. Nearly 58% of India's population depends on agriculture and allied activities for their livelihood, and 17-18% of GDP it contributes, still have low per capita income and increasing indebtedness.

The age-old practice of agriculture is found in India due to the marginal, small, and medium-scale farming of its majority rural population. The challenges in front of us are to abandon an unsustainable path and create a more evenly distributed growth path to protect the environment and future generations. As India adopted economic reforms from 1991 onwards based on globalisation policy, there has been a structural change in all sectors of the economy. On average, Indian economic growth has been between 7-9%, which is considered to be on the high side. Stable societies must have minimum ecological disruption and practice maximum conservation. Agriculture's sustainability is impossible to achieve as the population grows and economic growth accelerates.

The linkage between agriculture and industries and service sectors will go a long way towards ensuring the sustainability of economic development in India. The edited book is mainly empirical evidence and policies adopted by the centre and state governments in promoting agriculture development. Most of the studies conducted by the authors of primary data analyses some of the research studies are at a macro level. Another important aspect of sustainability practice is farmers' behaviour to understand its importance and adopt it for future generations. The other dimension is how far climate change has affected this sector. The edited book presents an analysis of India's agricultural development experience in favour of a sustainable development concept. Each author has approached the selected topic with the broad theme of the area of its title. They have used reliable data sets over time to uncover many quantitative dimensions of agricultural development. I believe this edited volume contributions read together make a positive contribution to the existing empirical literature on sustainable agricultural development. I hope that the articles covered in this book will help to strengthen agriculture development, policies, and strategies in coming years in the form of a future policy framework. I sincerely hope this edited book will appeal to those in the area of academic research and the non-academic community, as well as policy makers and readers interested in the deep issue of agricultural development.

Prof. Manjappa D Hosamane
Former Vice-Chancellor, Vijayanagar Sri Krishnadevaraya University, Bellary, Visiting Professor of Planning Commission Chair,
University of Mysore, Mysore

Book Introduction

Agriculture plays a crucial role in ensuring food security while also accounting for a significant share of India's Gross Domestic Product (GDP). It engages almost two-thirds of the workforce in gainful employment. Several industries such as sugar, textiles, jute, food and milk processing etc. depend on agricultural production for their requirement of raw materials. On account of its close linkages with other economic sectors, agricultural growth has a multiplier effect on the entire economy.

Sustainable agriculture can be defined in many ways, but it seeks ultimately to sustain farmers, resources and communities by promoting farming practices and methods that are profitable, environmentally sound and good for communities. Sustainable agriculture fits into and complements modern agriculture. It rewards the true values of producers and their products. It draws and learns from organic farming. It works on farms and ranches large and small, harnessing new technologies and renewing the best practices of the past.

Sustainable agriculture, simply defined, is an approach to agriculture that focuses on producing food in a way that does not degrade the environment and contributes to the livelihood of communities. However, this simple statement conveys a complex concept: that agriculture must balance production, environmental, and community development goals. The 1990 Farm Bill1 states that the term sustainable agriculture refers to "an integrated system of plant and animal production practices having a site-specific application that will over the long term:

- Satisfy human food and fiber needs.
- Enhance the environmental quality and natural resource base upon which the agricultural economy depends.
- Make the most efficient use of non- renewable resources and on-farm resources and integrate, where appropriate, natural biological cycles and controls.
- Sustain the economic viability of farm operations.
- Enhance the quality of life for farmers and society as a whole.

Overall, the book is crafted to bring out the expert opinions of different authors and is compiled and edited based on their scope.

Sustainable agriculture should not be an either-or proposition, such that a farm either is or is not sustainable. Rather, sustainable agriculture encompasses many principles and practices that benefit growers, their farm, their community, and the environment. The economic, environmental, and social goals of sustainable agriculture can serve as a useful yardstick for measuring a farm's performance and progress over time. This approach makes sustainable agriculture relevant to all farmers because it can be applied to farms of every size and type.

Sustainable agriculture aims to ensure the following elements,
- Produce safe and healthy food
- Conserve natural resources
- Ensure economic viability
- Deliver services for the ecosystem
- Manage the countryside
- Improve quality of life in farming areas
- Ensure animal welfare

The strategy of sustainable agriculture is generally concerned with the need for agricultural practices to be economically viable, to meet human needs for food, to be environmentally positive, and to concern with the quality of life. Sustainable development relies on economic, social and environmental foundations in the framework of coordinated worldwide governance. Feedback concerning the various economic, social and environmental policies already implemented must be obtained. The government must assume their responsibilities as a driving force in the field of sustainable development.

The strategy must be a facilitator of public opinion and policies in order to change consumption and investment behaviours. The strategy hinges on measures that take the main challenges into account, transverse measures, appropriate funding, the involvement of all the parties concerned, and the efficient implementation and monitoring of political decisions. The main directives of the strategy are the promotion and protection of basic rights, solidarity within and between generations, the guarantee of an open and democratic society, the participation of citizens, companies and the social partners, the coherence and integration of policies, use of the best available knowledge, the precautionary principle, and the 'polluter pays' principle.

Structure of the Book

The first chapter by Dr. Jayasheela and R. Sumithra deals with "Eradication of Hunger and Poverty through Sustainable Agriculture: An overview". This paper examines the hunger and nutrition situation prevailing in India and focused on the goals of an eradication of hunger and poverty through the sustainable agriculture.

The second chapter by Dr. Vinod Kumar S, and Dr. M Madhumathi envisions the "Diversification of Cropping Pattern and Sustainable Agricultural Development in India". The chapter gives the importance of Agricultural and allied sectors as one of the prominent sources for Indian economy. The implementation of sustainable agriculture practices can improvise cultivating which can cope with the ongoing demands and can also be one of the best alternatives for modern technologies which can tamper ecological security and imbalance the environment.

The third chapter by Manjunatha S Tyalagadi and Mahantesh Radderatti deals with "Recent Floods and associated Atmospheric and Oceanic factors in Indian Region: A Case study in Climate Change Scenario". The study aims in identify the most salient features during the flood infected periods in India. The study envisages the Datasets pertaining to Atmospheric and Oceanic conditions from NCEP which was utilized to reanalysis the Mean Monthly, Anomalies, Long term average/climatology maps. The results obtained in the study aids in forecasting the natural calamities to prevent major loss.

The fourth chapter by Jagadeesha B deals with "Farmers Attitude Towards Sustainable Agricultural Practices". The study highlights the importance of attitude towards building agronomical entrepreneurs. The information analysis was obtained from the application of the attitude scale toward sustainable agriculture which in turn developed the balanced environment and natural resources. The study pointed in the fact that higher the socioeconomic status and the greater knowledge access will lead to the greater sustainable agricultural practices.

The fifth chapter by Dr R Nagabhushan deals with "Sustainable-Food and Agriculture: Points to Ponder".The study highlighted the improvisation of agriculture comes with different factors such as the social and environmental costs, including water scarcity, soil degradation, ecosystem stress, biodiversity loss, depleting fish stocks and forest cover, and high levels of greenhouse gas emissions. These factors play immense important role to nurture and adapt sustainable lands to feed the upcoming generations.

The sixth chapter by Premakumari. L deals with "Sustainable Agricultural Systems and Practices in Indian states like Sikkim and Andhra Pradesh which are leading the path to adapt sustainable agricultural practices. The study also envisions on the scaling up context specific SAPSs which emerges to be crucial components to increase productivities and stand to benefit the most from the transition.

The seventh chapter by Dr.Siddappaji, D and Dr. Anand, C deals with "An Analysis of Sustainable Development of Agriculture Sector in Karnataka: An Overview". The chapter explored the sustainable agriculture farming in accordance with food and textile needs, without compromising the ability for future generations to meet their needs. It also states the sustainable agriculture is one of the ancient, primary occupations of Karnataka. It is the main source of livelihood for many people. It is the backbone of our state economy. Economic progress depends on agriculture. It provides to food for the people and raw materials to industries.

The eighth chapter by Girish M.C deals with "An Analysis of Environmental Impact on Agricultural Production in India – A Special Reference to Hassan District". The study observed that the environmental impact of agriculture varies based on the wide variety of agricultural practices employed around the world. Ultimately, the environmental impact depends on the production practices of the system used by farmers. The connection between emissions into the environment and the farming system is indirect, as it also depends on other climate variables such as rainfall and temperature. Weather is the condition of the atmosphere at a particular place and time.

The ninth chapter by Dr. Shashikumar, T.P. deals with "Agricultural Exports Policy in India: A Critical Appraisal". The study explains that India is currently ranked tenth amongst the major exporters globally as per WTO trade data. India's share in global exports of agriculture products has increased from 1 percent a few years ago, to 2.2 percent. The Agricultural Export Policy was introduced to increase and provide support to productivity, pre- and post-harvest management, value-addition and upgrade technology. The share of agricultural exports in India declined over a period of time. To make major reforms to the export policy for agriculture, India restructured from the Green Revolution Era and promoted Agricultural Export Policy to diversify the food and non-food agriculture base to emerge as a leading player in the world in agricultural trade. The policy would increase the agricultural exports leading to stable growth in GDP, benefits for farmers, employment in rural areas, quality and scope for value addition and future market potential.

The tenth chapter by Rajesha B deals with "Organic Farming as an opportunity for Sustainable Agriculture Development in India". The study opined that Agriculture, and hence choices made by a farmer, has a significant impact on environmental sustainability. Presently, more than 98% of farmers in India follow conventional farming using chemical fertilizers, often to the detriment of the environment, yields, and personal health. They remain hesitant to adopt organic farming despite its promise of greater sustainability and profitability. This chapter provides an overview of organic farming opportunities for sustainable agriculture development in India.

The eleventh chapter by Dr. Kiran, S. P. and Dr. Manjuprasad, C. deals with "A Study of Sustainable Development and its Impact on Rural Development in India". The Rural development has been one of the most formidable and fundamental aspect of India's developmental efforts. Rural development has remained a priority item through successive five-year plans. The policies and programmes for rural development is a strategy for the improvement of socio-economic and political life of the people with special emphasis on the rural poor. Rural poverty needs to be targeted and ended to achieve many of the Sustainable Development Goals but most particularly goal.

The twelth chapter by Dr. Manjuprasad, C. deals with "The Role of Cooperatives in Sustainable Agricultural Development". The present chapter observed that Sustainability in agriculture means the land and resources that use for agriculture today should be handed over to the future generations in a sustainable form so that they can continue to practice agriculture and have food security. Cooperatives can contribute to all SDGs, both because they are involved in the very diverse economic sectors concerned, and because their impact contributes substantially to the global objectives pursued.

The thirteenth chapter by Dr. J. L. Banashankari deals with "A Study on Organic Farming in India". The study explained that, in recent years, awareness of the harmful effect of chemical-based fertilizers and pesticides on our health is on a rise. Conventional agriculture relies heavily on chemical fertilizers and toxic pesticides etc., which enter the food supply, penetrate the water sources, harm the livestock, deplete the soil and devastate natural eco-system. Efforts in evolving technologies which are eco-friendly are essential for sustainable development and one such technology which is eco-friendly is organic farming. Organic food is growing in popularity across the world.

The fourteenth chapter by Chandra Prasad Hosamani and Dr.Shashikumar,T.P deals with "Government Efforts in Promoting Sustainable Agriculture in India: A Study". The present research paper analyse the schemes which are framed towards development of sustainable agriculture in India. The present chapter also analyse the various schemes which are framed towards development of sustainable agriculture in India. Major schemes like National mission for sustainable agriculture, Rainfed area development scheme, Submission on Agro forestry, Soil Health Management and Paramparagat Krishi Vikas Yojana are few of the government schemes which are enabling for Sustainable agriculture development in India.

The fifteenth chapter by Dr. Raghavendra B.N.deals with "An analysis of Sustainable Agricultural Development in India". This paper analyses the issues related to sustainable agriculture development. There is hope for positive changes as many farmers are becoming open-minded and government is coming up with new initiatives.

The study suggests that problems largely institutional, structural and administrative for development in sustainable agricultural development in particular. The private sector will be the key source of investment capital along with non-governmental organizations (NGOs) and civil society organizations (which benefit from local and international private knowledge when implementing programmes) will also play a key role.

The sixteenth chapter by M.N.Madhura deals with "Rural Development and Sustainable Agriculture Development". The present study explained that agriculture is demographically the broadest economic segment and plays a major portion in the overall socio-economic fabric of India. Agriculture with its allied sectors is unquestionably the main livelihood provider in India. So, in the vast rural areas 69 percent of India's population lives in rural regions and 3/4 of the people making up these rural populations depend on agriculture and allied activities for their livelihoods. Rural communities must develop several non-farming activities coupled with agricultural systems (adapted to local geographical conditions) to become more resilient to economic shocks or environmental disturbances in the context of climate change. Rural areas should receive the same attention and opportunities from decision-makers, academics, and professionals regarding sustainable development policies and investments in infrastructure projects. Agenda 2030 could be achieved if sustainable rural development policies will be implemented in each country next to urban areas.

About the Editors

Dr. Mahesha M., currently working as Professor at Department of Economics and Cooperation, University of Mysore, Mysuru. He did his Graduation, Post-graduation and Ph.D. degree from University of Mysore. He has twenty-eight years of teaching and research experience. He taught economic theory, industrial economics, mathematical and statistical economics and econometrics for post-graduate students. His areas of research include development economics, productivity and efficiency analysis, tourism economics and agriculture economics. He published more than seventy articles in national and international journals, presented around fifty papers in national and international conferences/seminars. He has delivered more than eighty invited lectures at workshops, symposia and conferences. He has organized six national and state level workshops on econometrics.

Dr. Shashikumar, T.P., Assistant Professor of Economics, has completed MA Economics in 2007 and M.Phil in 2008 from the University of Mysore. He has cleared UGC NET in 2012. He has done his Ph.D in 2016 from University of Mysore. His teaching expertise are in the field of Macroeconomics, Quantitative methods and international trade. His area of research includes development economics, agriculture economics, micro-finance and rural development. He has more than 12 years of teaching and research experience. He has published significant numbers of articles in various journals and edited books. He is the Co-ordinator, Audio section, Karnataka State Open University, Mysuru. He has conducted workshop and involved in many academic events. Presently, five doctoral students pursuing research under his supervision.

Dr. Manjunatha, C. S., Assistant Professor in Geography, has completed M.Sc. Geography in 2009 from the University of Mysore, Mysuru. He has cleared the SLET and NET in 2012. He has done his Ph.D. in 2020 from University of Mysore, Mysuru. He started his teaching profession as Assistant Professor at Karnataka State Open University from 2012. Currently, he is serving as Chairman, Department of Geography, Karnataka State Open University, Mysuru. He has published many research articles in national and international peer reviewed journals and many articles in edited books and conference proceedings.

X

Chapter 1

ISBN: 978-81-948672-7-2

Eradication of Hunger and Poverty through Sustainable Agriculture: An overview

DR. JAYASHEELA
Professor
Department of Studies and Research in Economics
Tumkur University, Tumkur-572103, Karnataka, India

Email: jayasheela_mu@yahoo.com

R. SUMITHRA
Research Scholar
Department of Studies and Research in Economics
Tumkur University, Tumkur-572103, Karnataka, India

Abstract: This paper examines the hunger and nutrition situation prevailing in India and focused on the goals of an eradication of hunger and poverty through the sustainable agriculture. The sustainable agriculture is the tool of an achieved food security through increased production of agricultural products. Food and good nutrition are basic human needs and this is recognised in the first Millennium Development Goal (MDG) – the eradication of extreme poverty and hunger. With more than 200 million people food insecure (FAO, 2008), India is home to the largest number of hungry people in the world. Hunger and food security are closely related to poverty. The Global Hunger Index (GHI) 2010 ranks India at 67 out of 122 countries; whereas the '2012 Global Hunger Index' ranks it at 65 among at 79 countries. Similarly, malnutrition in India, especially Survey Report released by save the children on 19th July 2012, India is ranked at 112 among the 141 nations as regards child development index. And there are disparities across various sections of the society and states. The collection of data which is collected from secondary sources of published articles, journals, government reports and annual reports etc. The present paper focused on an eradication of hunger and poverty to achieve food security and improved nutrition and promote sustainable agriculture. In this study, we study whether sustainable agriculture is an effective practice to eradicating the hunger and poverty in India?

Keywords: India, Sustainable Agriculture, Practices, Global Hunger Index, State India Hunger Index

Introduction

India is home to the largest number hungry people in the world. The global hunger index (GHI) 2020 ranks India at 101 out of 116 countries. Similarly, malnutrition in India, especially among children and women, is widespread, acute and even alarming. Poverty is the main cause of hunger and malnutrition. Government has initiated various measures to overcome hunger and malnutrition, but they are not so effectively implemented.

The chronic hunger and malnutrition in developing countries is the major challenge for policy planners and international funding agencies. This chapter discusses the sustainable agricultural development in India and it's exploring the possibilities through the achievement of eradication of hunger and poverty.

Sustainable Development

Sustainable development is defined as the development that meets the needs of present without compromising the ability of future generations to meet their own needs". The concept of sustainable development has two dimensions – to make better and to maintain and the primary focus of sustainability is on the issue of intergenerational equity, which implies equal availability of options in terms of human well-being or production prospects to future generation as compared to the present one. Sustainable development is a multidimensional concept with three interacting angles for natural resource management: ecological security, economic efficiency and social equity.

Sustainable Agriculture
History

Agriculture has changed dramatically since the end of World War II. Food and fibre productivity has soared due to new technologies, mechanizations, increased chemical use, specialization and government policies that favoured maximizing production and reducing food prices.

These changes have allowed fewer farmers to produce more food and fibre at lower prices. Although these developments have had many positive effects and reduced many risks in farming, they also have significant costs. Prominent among these are topsoil depletion, ground water contamination, air pollution, green house gas emission, the decline of family farms, neglect of the living and working conditions of farm labourers, new threats to human health and safety due to the spread of new pathogens, economic concentration in food and agricultural industries, and disintegration of rural communities.

A growing movement has emerged during the past four decades to question the necessity of these high costs and to offer innovative alternatives. Today this movement for sustainable agriculture is garnering increasing support and acceptance within our food production systems. Sustainable agricultural integrates three main goals- environment health, economic profitability, and social equity.

Sustainable Agricultural in India

In the last few decades, India has achieved food security through increased production of rice and wheat. Still, attaining nutrition security remains a challenge. Around 22 percent of India's adult population (15-49 years) is undernourished and more than 58 percent of Indian children up to 5 years are anaemic.

While the green revolution's promotion of high – yielding varieties of seeds and fertilizers did solve food-grains shortages, its drawback is now visible in the form of degraded land, soil, and water quality as farmers declining incomes due to a high dependency on external inputs.

Climate change poses another serious threat to Indian agriculture. Which is largely rainfed and fundamentally dependent on climatic stability. With the projected 1.5-degree Celsius increase in the planet's average atmospheric temperature and greater variability in summer monsoon precipitation, risks to food security, livelihoods, water supply, and human well-being are bound to increase.

REVIEW OF LITERATURE

Ahmad, Rais, Aligarh Muslim University, 2013 "Sustainable Agriculture Development in India: A Case Study of Uttar Pradesh". Their view on agriculture is a critical sector of the Indian economy. It forms the backbone of development in the country. An average Indian still spends almost half of total expenditure on food, while roughly half of India's work force is engaged in agriculture for its livelihood. Agriculture is source of society of livelihood and food security for a vast majority of low income, poor and vulnerable sections of society. The study is entirely based on secondary sources of data collected from different official documents and websites of government of India and Uttar Pradesh.

Adedokun. M.O.(Ph.D) Ekiti State University (2018), "Fighting Hunger and Poverty for Child's Sustainability: A case study of Ibadan, Metropolis, Oyo State, Nigeria, their view on the study showed that children in Ibadan, Oyo state Nigeria are denied rights to education, good health care, adequate nutrition and safe water among others due to poverty and for sustainable growth to the encouraged all aspects of children's lives such as health, child protection, nutrition as well as social inclusion should be taken care of. The study recommended that the government should put in place effective policies capable of ending poverty and promoting sustainable growth and the government should make polies that would enhance food security.

Poornima Menon, Anil Deolalikar, Anjor Bhaskar, Washington DC (2009) "India State hunger index Comparisons of hunger across states" N.C Saxena, (January 2011) "Hunger, Under-Nutrition and Food Security in India". This paper examines the hunger and nutrition situation prevailing in India and suggests policy measures for ensuring adequate food security at the household level, particularly for marginalised groups, destitute people, women and children.

HUNGER AND MALNUTRITION IN INDIA: STATUS, CAUSE AND CURSE-

National Situationer- Association of Voluntary Agencies for Rural Development (AVARD)5(FF), Institutional Area, Deen Dayal Upadhyay Marg, New Delhi – 110002.

Montek Ahluwalia, (January 2005), "Reducing Poverty and Hunger in India: The Role of Agriculture. Their view on India's strategy for reducing poverty and hunger has always placed a great deal of importance on the agriculture sector, reflecting the fact that 70 percentage of the population live in rural areas and the overwhelming majority of them depend upon agricultural as their primary source of income.

Bharat Kumar Devda, (February 2001), "Reducing food poverty with sustainable agriculture: A summary of new evidence.

Objectives of the Study

1. To Know the study of an Overview of sustainable agricultural Practices in India
2. To understand the how sustainable eradicating the hunger and poverty in India
3. To review the current status of global hunger index and severity rate in India
4. To analyse the study of global hunger index and state wise India hunger index number

Methodology

The study is entirely based on secondary sources of data, which is collected from different official documents and websites of government of India, published articles, journals etc.

An Overview of sustainable agricultural practices in India

Sustainable agriculture is a wide notion that encompasses a variety of methods to get the stage of sustainable agriculture, the following procedures are used.

Organic Farming

Organic Farming is a production system that prohibits the use of synthetically produced agro-inputs (Fertilisers and pesticides) instead, it relies on organic material (such as crop residues, animal residues, legumes, bio-pesticides) for "maintaining soil productivity and fertility and managing pests under conditions of sustainable of sustainable natural resources and a healthy environment".

Nearly, 2.8 million of area under certified organic farming as of March 2020, 1.9 million registered and certified farmers as of March 2020. Cereals, millets, cotton, fruits, vegetables, and more – are being cultivated using organic farming.

Natural Farming

Natural farming in the Indian context is a local low-input climate resilient farming system that advocates the complete elimination of synthetic chemical agro- inputs. Instead, it encourages farmers to use low -cost, locally sourced inputs such as natural mixtures made using cow dung, cow urine, jaggery, pulse flour. it also encourages mulch, crop covers, and symbiotic intercropping to stimulate the soil's microbial activities. Natural farming's main emphasis is on "enhanced soil conditions by managing organic matter and soil biological activity. Diversification of genetic resources, enhanced biomass recycling; and enhanced biological interactions. Altogether, 6,52,000 ha of area under natural farming across   Andhra Pradesh, as of November 2020, 6,377 ha area under natural farming in Himachal Pradesh as of March 2021. Nearly, 6,00,000 Farmers enrolled in the Andhra Pradesh state programme for natural farming, as of November 2020, 1,16,700 Farmers are practising natural farming under the Himachal Pradesh's Prakritik Kheti Khushal Kisan Yojana as March 2021. Cereals, millets and cotton to fruits, vegetables, and spices ate cultivated under natural farming.

System of Rice Intensification

The system of Rice Intensification is a climate-smart agroecological approach for increasing rice and other crops productivity by changing the management of the plant, soil, water and nutrients. It is based on the main four principles that interact with each other: i), early, quick and healthy plant establishment, ii), reduced plant density; iii) improved soil conditions through enhancing soil organic matter; iii) reduced and controlled water application. 3 million area under SRI across different states in India, beyond rice the SRI principles are also being applied to wheat, sugarcane, and pulses.

Agroforestry

Agroforestry describes traditional and modern land-use systems where woody perennials (trees, shrubs, bamboos, palms) are in various spatial or temporal arrangements. 25 million area is under agroforestry across 15 agroclimatic zones. Less than 5 million farmers practise agroforestry across India.

Precision Farming (PF)

PF is an approach to farm management that uses information technology to ensure that the crops and soil receive exactly what they need for optimum health and productivity, rather than applying similar inputs across the entire field, the approach aims to manage and distribute them on a site-specific basis to maximise long-term benefits and prevent waste. Altogether, 9.2 million has been covered under precise micro irrigation techniques – drip and sprinkler, the two most widespread PF techniques in India.

Conservation Agriculture (CA)

CA is an ecosystem approach to agricultural land management based on three interlinked principles: i) minimum disturbance to soil through reduced tillage, ii) permanent maintenance of soil mulch by retaining crop residues, iii) diversification of cropping through crop rotation and intercropping.

Crop Rotation and Intercropping

Crop Rotation is the practice of planting two or more crops sequentially on the same plot of land to improve soil health, optimise nutrients and combat pest and weed pressure. Simple rotation may involve two or three crop, while a complex rotation may incorporate a dozen or more. Intercropping is the growing of two or more crops simultaneously in the same field and can be of various types – mixed, row, strip, and relay intercropping.

Cover Crops and Mulching

Cover crops are crops planted to cover the soil rather than to be harvested, they can be rotated with other crops or intercropped and also grown in between cultivation seasons to control soil erosion, and organic matter to the soil, supplying nitrogen, controlling weeds and fighting insects. Mulching is the practice of covering the soil surface with organic materials. 1.9 million ha under cover crops, which includes plantations having leguminous cover crops.

Integrated Pest Management

IPM system consists of using suitable techniques and methods in a compatible manner to maintain pest populations at levels below those causing economically unacceptable damage or loss. It combines cultural, biological and chemical measures to provide a cost-effective, environmentally-sound and socially acceptable method of controlling disease, insects, weeds, etc.

Vermicomposting

Vermicomposting is a biotechnological composting process that uses certain earthworms to enhance the process of biomass waste conversion to produce good-quality compost. The resultant product is a stabilised uniformly sized substance with a characteristic earthy appearance known as vermicast.

Biodynamic Farming

The biodynamic farming system mainly works on the relationship between plant growth and cosmic rhythms and emphasises the importance of maintaining sustainable soil fertility.

Contour Farming

Contour farming is ploughing and planting along a contour-across the slop rather than up and down furrows are ploughed perpendicular rather than parallel to the slope.

Integrated Farming System

IFS can be describing as a judicious mix and positive interaction between two or more components – such as horticulture, crops, livestock, aquaculture, poultry, apiculture, and mushroom cultivation.

Rainwater harvesting

Collects, conveys and stores the rainfall in an area for beneficial purposes. It is done by storing rainwater on the surface for future use and through recharge to groundwater.

Floating Farming

Floating farming is a way of producing food in areas that are waterlogged for long periods. It is mainly aimed at adapting cultivation to increased or prolonged flooding. The system uses floating beds of water hyacinth, mud, and bamboo.

Permaculture

Permaculture is described as "consciously designed landscapes. Which mimic the patterns and relationships found in nature while yielding an abundance of food, fibre, and energy for provision of local needs.

Eradication of hunger and poverty through sustainable agriculture development

For eradication of hunger, availability of food grains is one of the important components. Plan, agricultural policy was focused on increasing yields, productivity and general food availability in the country. The green revolution was launched to increase productivity and production. Impact of the green revolution show that predominant focus on the increased production of rice and wheat-staple grains consisting mainly of carbohydrates, protein and a few other nutrients are essential to meet human nutritional requirements. This push to concentrate on a few staple crops may be a contributory factor to the simplified diets that continued under nutrition in South Asia and widespread nutritional deficiencies. These policies could not achieve improved food and nutrition security as the preference of child malnutrition is the highest in India. While the green revolution improved food productivity significantly, the role of the health care, childcare and diverse and quality foods for household food and nutrition security was less emphasized. Agriculture provides sources of income but not improvement in child nutrition.

Hunger and Malnutrition

Hunger: The problem of hunger is complex, and different terms are used to describe its various forms. Hunger is usually understood to refer to the discuss associated with a lack of sufficient calories.

Undernutrition: Goes beyond calories and signifies deficiencies in any or all of the following: energy, protein, essential vitamins and minerals. Undernourishment is the result of inadequate intake of food in terms of either quantity or quality, poor utilization of nutrients due to infections or illness Ness or a combination of these factors.

Malnutrition: Refers more broadly to both undernutrition and overnutrition problems caused by unbalanced diets, such as consuming too many calories in relation to requirements with or without low intake of malnutrition-rich foods.

Hunger Index Number in India

Food and good nutrition are basic human needs, and this is recognized in the first millennium development goal- the eradication of extreme poverty and hunger. The global hunger index is one approach to measuring and tracking progress on hunger and enabling widespread discussion about the reasons for and consequences of hunger. Although hunger is most directly manifested food intake, over time inadequate food intake and a poor diet. Especially in combination with low birth weights and high rates of infection. Can result in stunted and underweight children. The ranking of nations on the basis their index scores has been a powerful tool to help focus attention on hunger, especially for countries like India that underperform on hunger and malnutrition relative to their income levels.

Table: 1. Global Hunger Index Score By 2021 GHI index score at Global Wise

Rank	Country	2000	2006	2012	2021
2021 GHI scores less than 5, collectively ranked 1-18	Belarus	< 5	< 5	< 5	< 5
	Bosnia &Herzegovina	9.3	6.7	< 5	< 5
	Brazil	11.5	7.4	5.5	< 5
	Chile	< 5	< 5	< 5	< 5
	China	13.3	9.0	< 5	< 5
	Croatia	< 5	< 5	< 5	< 5
	Cuba	< 5	< 5	< 5	< 5
	Estonia	< 5	< 5	< 5	< 5
	Kuwait	< 5	< 5	< 5	< 5
	Latvia	5.5	< 5	< 5	< 5
	Lithonia	< 5	< 5	< 5	< 5
	Montenegro	-	6.5	< 5	< 5
	North Macedonia	7.5	7.7	< 5	< 5
	Romania	7.9	5.9	5.0	< 5
	Serbia	-	5.9	5.0	< 5
	Slovakia	6.0	5.3	< 5	< 5
	Turkey	10.2	6.5	5.0	< 5
	Uruguay	7.4	6.7	5.0	< 5
19	Argentina	6.4	5.6	5.2	5.3
19	Costa Rice	7.0	5.5	< 5	5.3
21	Uzbekistan	24.3	16.0	9.5	5.9
22	Tunisia	10.3	7.8	7.0	6.0
23	Bulgaria	8.6	8.1	7.8	6.1
23	Mongolia	30.2	23.4	12.8	6.1
25	Albania	20.7	15.9	8.8	6.2
61	Vietnam	26.3	21.8	16.0	13.6
62	Ecuador	19.7	18.9	12.8	14.0
62	Nicaragua	22.3	17.4	14.9	14.0

64	Ghana	18.4	22.0	17.9	14.9
65	Sri Lanka	21.9	20.0	20.6	16.0
66	Sengal	34.0	24,1	19.2	16.3
67	Gabon	21.0	20.2	18.6	16.6
68	Philippines	25.0	20.4	20.5	16.8
69	Cambodia	41.1	27.1	24.2	17.0
69	Eswatini	24.5	23.2	21.8	17.0
72	Myanmur	39.8	31.6	22.9	17.5
73	Indonesia	26.1	29.5	23.0	18.0
74	Cameroon	35.7	30.9	23.1	18.6
75	Solomon Island	20.0	18.2	20.2	18.8
76	Bangladesh	34.0	30.9	28.6	19.1
76	Nepal	37.4	31.9	23.1	19.1
78	LAO pdr	44.1	24.6	25.7	19.6
79	Guatemala	28.4	24.6	22.0	19.6
*	Tajikistan	-	-	-	10-19.9*
80	Namibia	25.3	25.8	26.6	21.3
81	Malawi	43.1	33.5	26.2	22.2
81	Benin	34.0	27.7	24.0	22.2
82	Venezuela	14.6	11.2	7.4	22.3
84	Cotedlvoire	33.3	37.1	30.0	22.6
85	Mauritania	31.9	28.9	23.63	22.8
86	Iraq	23.9	31.2	27.5	23.0
87	Kenya	36.9	26.2	25.4	23.2
88	Botswana	36.7	36.5	24.3	23.7
89	Togo	26.7	43.4	25.3	24.1
90	Ethiopia	39.1	35..8	33.5	24.5

91	Burkina Faso	53.3	36.8	29.7	24.7
92	Mali	44.9	33.1	24.8	24.7
93	Pakistan	41.7	33.6	32.1	24.7
92	Tanzania	36.7	33.6	29.1	25.1
92	Sudan	-	-	29.8	25.2
95	Korea	39.5	33.1	29.1	26.0
96	Angola	65.0	46.9	27.8	26.4
97	Rwanda	49.3	38.3	31.0	27.4
98	Djibouti	44.3	36.9	35.4	27.4
99	Lesotho	32.5	29.6	24.6	27.5
101	**India**	**38.8**	**37.4**	**28.8**	**27.8**
102	Papua New Guinea	33.6	30.3	33.7	28.3
103	Afganistan	50.9	42.7	34.3	28.3
25	Russian Federation	10.1	7.1	6.4	6.2
27	Georgia	12.3	8.8	< 5	6.3
28	Kazakhstan	11.2	12.3	8.1	6.4
29	Saudi arabia	11.0	12.1	8.2	6.8
29	Ukraine	13.0	7.1	6.9	6.8
32	Armania	19.3	13.3	10.4	7.2
31	Algeria	14.5	11.7	8.9	6.9
33	Azerbaijan	25.0	15.9	10.6	7.5
33	Paraguay	11.7	11.6	9.5	7.5
35	Iran	13.5	8.9	8.1	7.7
36	Dominican Republic	15.1	13.2	10.2	8.0

36	Peru	20.6	16.4	9.2	8.0
38	Jordan	10.8	8.1	8.5	8.3
39	Mexico	10.2	8.6	7.8	8.5
40	Fiji	9.6	9.0	8.1	8.6
40	Jamaica	8.6	9.0	9.1	8.6
40	Kyrgyzstan	18.3	13.9	11.7	8.6
43	Morocco	15.5	17.5	9.6	8.8
44	Colombia	10.9	11.4	9.3	8.9
44	El Salvador	14.7	12.0	10.4	8.9
44	Panama	18.7	15.0	10.1	8.9
44	Trinida & Tobago	11.0	11.3	10.8	8.9
*	Moldova	-	-	-	-
50	Suriname	15.1	11.4	10.4	10.4
51	Guyana	17.1	15.6	12.1	10.7
53	Coba Verda	15.4	11.9	12.3	10.8
53	Thailand	18.5	12.3	12.4	11.7
54	Mauritus	15.2	14.0	13.0	12.2
55	Oman	14.7	13.8	11.6	12.3
56	Egypt	16.3	14.4	15.2	12.5
57	Bolivia	27.8	23.3	15.6	12.7
58	Honduras	21.8	19.6	13.8	12.8
58	Malaysia	15.4	13.7	12.4	12.8
60	South Africa	18.1	17.6	12.7	12.9

Table 1. Indicates that India is home to the largest number hungry people in the world. The global hunger index (GHI) 2020 ranks India at 101 out of 116 countries. Similarly, malnutrition in India, especially among children and women, is widespread, acute and even alarming.

Poverty is the main cause of hunger and malnutrition. Government has initiated various measures to overcome hunger and malnutrition, but they are not so effectively implemented.

Table: 2 Measuring Scale of GHI index number (2021)

GHI Severity scale of India				
Less than equal to 9.9	10.0-19.9	20.0-34.9	35.0-49.9	Less than 50.0
Less than 50.0	Moderate	Serious India at 27.8	Alarming	Extremely Alarming

(**Source**: (https://www.globalhungerindex.org/about.html)

Table 2 Indicates that according to GHI 2020 Index India at 101 out of 116 countries. India's hunger situation at 27.8 percentage, so it was alarming situation for India at globally.

Table: 3 Measuring Scale of GHI index number (2021)

STATE	Prevalence of calories under nourishment (%)	Proportion of Underweight among children <5years (%)	Under-five mortality rate (deaths per hundred)	India State Hunger Index Score	India State Hunger Rank India
Punjab	11.1	24.6	5.2	13.63	1
Kerala	28.6	22.7	1.6	17.63	2
Andhra Pradesh	19.6	32.7	6.3	19.53	3
Assam	14.6	36.4	8.5	19.83	4
Haryana	15.1	39.7	5.2	20.00	5

Tamil Nadu	29.1	30.0	3.5	20.87	6
Rajasthan	14.0	40.0	8.5	20.97	8
West Bengal	18.5	38.5	5.9	20.97	8
Uttar Pradesh	14.5	42.3	9.6	22.13	9
Maharashtra	27.0	36.7	4.7	22.80	10
Karnataka	28.1	37.6	5.5	23.73	11
Orissa	21.4	40.9	9.1	23.80	12
Gujarat	23.3	44.7	6.1	24.70	13
Chhattisgarh	23.3	47.6	9.0	26.63	14
Bihar	17.3	56.6	8.5	27.30	15
Jharkhand	19.6	57.1	9.3	28.67	16
Madhya Pradesh	23.4	59.1	9.4	30.87	17

Note: The India state hunger index represents the index calculated using a calories undernourishment cut off of 1.632 Kcals per person per day to allow for comparison of the India state hunger index with the global hunger index 2008

Source: India State Hunger Index

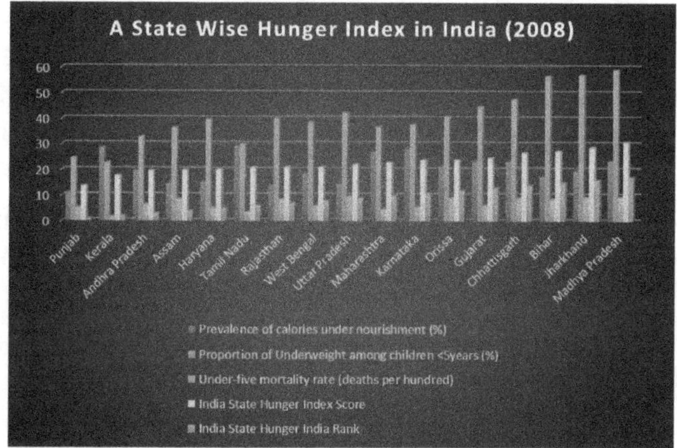

Source: India State Hunger Index

Above the Figure 1.2.4 indicates that, Madhya Pradesh has the highest rate of hunger ratio in the state wise at alarming situation and followed by Jharkhand, Bihar, Chhattisgarh Gujarat and Orissa   at serious condition of hunger and malnutrition rate in India.

Table: 3 GHI Severity scale of India State Hunger Index

GHI Severity scale of India State Hunger Index				
Less than equal to 9.9	10.0-19.9	20.0-34.9	35.0-49.9	Less than 50.0
Low	Moderate	Serious	Alarming	Extremely Alarming
	Punjab, Kerala, Andhra Pradesh, Assam	Haryana, Tamil Nadu, Rajasthan,West Bengal, Uttar Pradesh, Maharastra Karnataka Orissa Gujarat Chhattisgarh Bihar Jharkhand	Madhya Pradesh	

The table 3 indicates that, there are four categories low, moderator, serious, alarming and extremely alarming of Hunger Index in state wise in India, with only one state Madhya Pradesh falling in the alarming category, remaining states Haryana, Tamil Nadu, Rajasthan, West Bengal, UP, Maharashtra, Orissa etc, falling in the serious Condition and Punjab, Kerala, Andhra Pradesh, Assam in the Moderate Condition According to the report 2008 India State Hunger Index.

Conclusion

Finally, concluded that without effective reliable estimate, any future policy proposals for eradicating the hunger and poverty through the achievement of sustainable agriculture are not going to be effective. It is important that the agriculture is a major component to overcome the problems of food insecurity, under nutrition.

References

Ahmad, M.R. (2013). Sustainable Agriculture Development In India: A Case Study Of Uttar Pradesh.

Adedokun (2018). Fighting Hunger and Poverty for Child' s Sustainability: A Case Study of Ibadan, Metropolis, Oyo State, Nigeria.

Saxena, N.C. (2011). Hunger, Under-Nutrition and Food Security in India.

Ahluwalia, M.S. (2005). Reducing poverty and hunger in India: the role of agriculture.

Devda, B.K. (2001). Reducing food poverty with sustainable agriculture: A summary of new evidence.

(https://www.globalhungerindex.org/about.html)

www.statista.com

Chapter 2   **ISBN: 978-81-948672-7-2**

Diversification of Cropping Pattern and Sustainable Agricultural Development in India

Dr. Vinod Kumar. S
Lecturer in Economics
Maharani Cluster University
Bengaluru

Email:ymnvnd@gmail.com

Dr. M. Madhumathi
Professor
Department of Economics
and Director of Humanities and Liberal Arts, Maharani Cluster University, Bengaluru

Abstract: Agricultural and allied sectors are considered to be the mainstay of the Indian economy. They are the important source of raw material and generate demand for many industrial products, particularly fertilizers, pesticides, agricultural implements and a variety of consumer goods. The agriculture sector has performed impressively in terms of increasing productivity and intensity of cultivation. Unlike in the past, crisis situations like drought or flood are now considerably well managed without any panic or large- scale imports. On the other hand Sustainable agriculture plays on important role in raising crops for greater human utility through utilization of resources with better efficiency without disturbing, misbalancing or polluting the environment. But at the same time indiscriminate use of modem technology may endanger ecological security and imbalance the environment. The paper highlights the Diversification of cropping pattern, sustainable agricultural Development and trends in Area, Production and Productivity of Agricultural crops in India. To achieve sustainable agricultural development in India some policy measures are suggested.

Keywords: Crop Diversification, Sustainable Agricultural Development, Production and Productivity of Agricultural Crops

Introduction

India is the seventh largest and second most populous country in the world. Agricultural sector plays an important role in the economic development of India. It is a major source of livelihood for millions of people in India.

Approximately 54.6 percent of the total workforce depends on agriculture either directly or indirectly. The annual agricultural growth rate was 4 percent and total food grain production stood at 308.65 million ton in 2020-21. Agricultural sector contributed 17.8 percent to India's GDP.

Agriculture and allied' industry are further divided into several segments, namely, horticulture and its allied sectors (including fruits and vegetables, flowers, plantation crops, spices, aromatic and medicinal plants); fisheries sector; animal husbandry and livestock; and sericulture. India's varied agro-climatic conditions are highly favorable for the growth of large number of horticultural crops, which occupy around 10 per cent of gross cropped area of the country producing 160.75 million tons.

India is the second largest producer of fruits and vegetables in the world. It is also the second largest producer of flowers after China. It is also a leading producer, consumer and exporter of spices and plantation crops like tea, coffee, etc. while sericulture is an agro-based cottage industry. India is ranked as the second major raw silk producer in the world.

Review of Literature

This study covers the various, books, articles, empirical papers, journals, moreover very helpful source of internet.

Vani and Vyasulu (1996),Sandeep (2002),Wadear (2003),BrijBala and S.D Sharma(2005), have discussed about trends in cropping pattern and their results in most of the cases indicated that a positive growth rate in the production and yield of selected agriculture crops and a negative growth rate in area under selected agriculture crops.

Ashok Gulati, Pratap (2004), Santhosh Rathod, Joshi, et al, (2007) P Parthasarathy Rao, PK Joshi, Shravan Kumar and Kavery Ganguly ((2008) Brajesh Jha, Nitesh Kumar and Biswajit Mohanty (2009) they analysed agricultural diversification in different regions observed that, diversification of agriculture opens up opportunities for value addition in agriculture and enhanced income earning opportunities for the farming community.

Varghese (2004) Manoj S. Kamat, Sanjay N. Tupe, Manasvi M. Kamat (2007) have discussed about trends in area, production and productivity and found that changes in cropping pattern may be due to various reasons and also found that Indian agriculture sector has witnessed Decreasing Returns to Scale after the introduction of economic reforms, indicating that the input availability is under strain during the same period. The present study tries to fill up the gap which was not worked in previous period.

Objectives
- To Study the Objectives and Determinants of Crop Diversification.
- To analyze the Diversification of cropping pattern and Sustainable Agricultural Development.
- To study the trends in area, production and yield of selected agricultural crops in India.
- To suggest the policy measures for sustainable agricultural development in India.

Hypotheses
There is a paradigm shift of agricultural land from food to non-food crops over the period of time in India.

Database and Methodology
The present study is based on secondary sources. The sources of secondary data have been collected from different reports such as various economic survey reports Directorate of Economics and Statistics, Ministry of Agriculture and Co-operation Government of India, Handbook of Statistics on Indian economy, RBI. The data has also been collected from various articles, journals, internet sources like krishiworld.com, kissanworld.com. Etc.

Appropriate statistical and econometrics tools have been used to obtain the results by using SPSS statistical packages.

Discussion

Cropping pattern

Cropping pattern refers to the acreage distribution of different crops in any one year in a given farm area such as a district, water agency, or farm. Thus, a change in a cropping pattern from one year to the next can occur by changing the relative acreage of existing crops, and/or by introducing new crops, and/or by cropping existing crops. The cropping pattern of a region reveals the proportion of area of land under different crops at a point of time, the rotation of crops and the area under double cropping. The cropping pattern changes in space and time. In fact, no cropping pattern can be good and ideal for all times to come.

The cropping systems of a region are decided by and large, by a number of soil and climatic parameters which determine overall agro-ecological setting for nourishment and appropriateness of a crop or a set of crops for cultivation. Nevertheless, at the farmer's level, potential productivity and monetary benefits act as guiding principles while opting for a particular crop/cropping system. These decisions with respect to choose of crops and cropping systems are further narrowed down under the influence of several other forces related to infrastructure facilities, socio-economic factors and technological developments, all operating interactively at micro-level.

The cropping pattern plays a vital role in determining the level of agricultural production and reflects the agricultural economy of an area/ region. A change or shift in cropping pattern implies a change in the proportion of area under different crops which depends, to a large extent, on the facilities available to raise crops in the given agro-climatic setting. The natural, social, economic and historical factors determine the cropping pattern of a region. The cropping pattern also changes in consonance with the government policies and technological innovations especially in agriculture. It is, however, pertinent to mention that in most of the areas it is the availability of water more than any other input which determines the nature of agricultural production.

Cultivators in regions and localities without sources of water-supply to supplement rainfall are constrained in their choices of cropping patterns and have to suffer enforced adjustments when the rains are abnormal, either in their timing or in their quantity. Hence the state of agriculture of a region depends vitally upon the security and flexibility provided by irrigation facilities.

Patterns of crop diversification:

Crop diversification means growing a variety of crops in an area, not just one. If one crop fails in a given year, the area can still survive. The growth and spread of predators are limited.

In relation to agricultural development, —diversification is probably one of the most frequently used terms in the recent decade. Traditionally, diversification was used more in the context of a subsistence kind of farming, wherein farmers grew many crops on their farm. In the recent decade, diversification is increasingly being used to describe, increase in area under high value crops.

In this perspective it is conducive to explain what exactly diversification is?

Diversification originated from the word —diverge, which means to move or extend in a different direction from a common point. In this sense diversification is the opposite of concentration, therefore, most of the techniques of measuring diversification actually measures concentration in the system. In economics, diversification refers to a situation in which decrease in the dominance of an activity, alternately results in depiction of increase in the share of many activities in the system. Extending the same notion to agriculture, it means increase in the share through withdrawal of a resource and to allocate it with other alternatives. Diversification is therefore measured with concentration ratios. The concentration indices however do not explain the alternate definition of agricultural diversification that is, increase in the share of high value crops in agriculture. The high value range of crops is definitely wider than fruits and vegetables. The study therefore measures diversification with the changes in the percent of non-food crops at the aggregate level. This will also contribute to the recent debate on food versus non-food crops in the country.

Diversification of Cropping Patterns refers to bringing about a desirable change in the existing cropping patterns towards more balanced cropping system. In agriculture, diversification refers to the addition of new crops or enterprises with or without the addition a shift from one crop or enterprise in a production system. This includes horizontal as well as vertical diversification of agriculture.

Horizontal Crop Diversification

Horizontal crop diversification stands for inclusion of more and varied crops in the cropping system, using multiple cropping techniques, rather than concentrating on repetition of few crops. One desirable aspect of such diversification is to promote those cropping patterns which lead to sustainable agricultural practices.

Vertical Crop Diversification

Vertical crop diversification stresses upon the development of allied sectors and shift of burden from cultivation to allied activities example, animal husbandry, horticulture, floriculture, food and fruit processing etc. For the purpose of the study, only horizontal diversification has been considered.

Objectives of Crop Diversification
Crop diversification targets to achieve the following objectives:

(i) Equilibrium in demand and supply of agricultural produce: As a response to the Green Revolution, the production of wheat, rice and coarse cereals has surged up in Punjab, Haryana and Western Uttar Pradesh, leading to the state of food sufficiency. However, with improvement in standard of living, the demand for varied produce e.g. pulses, oilseeds, fruits and vegetables has increased manifold. Thus, in order to streamline demand and supply, crop diversification is an effective strategy.

(ii) To ensure gainful returns to the farmers: For this purpose, diversification needs to be in favor of high value crops. When high value crops are included in the cropping system, net returns to the farmers are maximized.

(iii) To provide hedge against adverse agro-climatic conditions: Crop diversification can shift the risks associated with monsoon failure, floods and the like conditions. If one of the crops is affected, its loss can be offset by other crop. Some crops have better climatic resilience. These should be included in the cropping system.

(iv) To provide hedge against price fluctuations: Prices of agricultural produce are subject to the application of Cobweb model. As such diversification can provide a hedge against price fluctuation of select crops and ensure better returns to the farmers.

(v) To strengthen agro ecological system: Each crop requires different set of nutrients from the soil. If it is repeated time and again, soil fertility gets reduced and high intensity of fertilizers is required. Diversification ensures restoration of nutrients and hereby rejuvenation of the soil.

vi) Shifting additional burden from agriculture: By diversification of agriculture and development of allied sectors, additional burden can be shifted, making it remunerative to the persons involved.

(vii) Optimum utilization of resources: While selecting a particular crop, the opportunity cost of alternative crop shall always be considered by making cost-benefit analysis of the crops. If diversification is commensurate with such consideration, it leads to optimum utilization of resources.

(viii) Sustainable development of the farm sector: To ensure food and nutritional security to the large population, sustainability of farm sector is the necessity. With diversification of cropping patterns, soil health can be retained in rejuvenated form.

Determinants of Crop Diversification

Selection of a particular crop over other alternates is a complex problem, affected by multiplicity of factors. These factors include economic as well as non-economic considerations. Due to lack of application of knowledge, research and extension services in the farm sector, such a decision is seldom taken wisely.

Following are the factors determining the selection of a particular crop in the cropping system:

(i) Resource related factors like irrigation facility, expected rainfall, soil fertility etc.

(ii) Agricultural technology related factors like availability of quality seeds, fertilizers, irrigation technologies etc.

(iii) Infrastructural variables like logistics, warehousing, regulated markets etc.

(iv) Institutional variables like regulatory policies, research and extension facilities, pricing policy etc.

(v) Household related factors like number of family members, working members, food and fodder requirement etc.

(vi) Responsiveness of farmers to the Government policies and programs in this direction.

Sustainable Development in Indian Agriculture

Sustainable development means, development at present meets the needs of the present generation without compromising the ability of future generation to meet their own demand.

Sustainability in agriculture means the land and resources that use for agriculture today should be handed over to the future generations in a sustainable form so that they can continue to practice agriculture and have food security.

This means that we have to use lands, water resources, etc in such a manner that the future generations are also will be able to have sustainable development. Sustainable agriculture is the system of raising crops for greater human utility through utilization of resources with better efficiency without disturbing imbalancing or polluting the environment. Sustainable agriculture is ecologically sound, economically viable, and socially justifiable for the human.

Improving the Sustainability in Indian Agriculture

Agriculture

Diversification of land suitable for farming to non-farm uses should be prevented by legislation. The soil with diminishing biological potentials should be improved through the efficient adoption of the principle of restoration of ecology. Uncultivated land must be re-vegetated. Efforts must be taken towards arresting soil erosion, conservation of water and biological diversity. The consumption of mineral fertilizers and chemical pesticides must be reduced. The integrated pest management system involving crop rotation, green water potentials must be effectively used to save research and training to achieve sustainability in agriculture, think nationally but plan and act locally.

Forest

A Scheme of integrated forest management which can take care of the triple needs of conservation, community needs, and commercials needs will have to be developed.

Protecting Environment

Measures such as removal of subsidies on pesticides, development of pest resistant varieties and implementation of integrated pest management must be adopted by the Government.

Preserving Genetic Resources

A special programme is necessary for the collection, conservation, evaluation and enhancement of crop genetic resources directly related to the promotion of sustainable advances in crop productivity.

Other Needs

1. Removing subsidies which encourage excessive use of fossils fuels, irrigation water pesticides and excessive logging.
2. Acceleration of provision of sanitation and clean water, agricultural extension credit and research.

Aquaculture

Use of eco-system approach to the management of marine resources. No single satisfactory indicator is available to study the short run progress towards sustainable agricultural development, because it involves value judgements such as the level of genetic diversity necessary to ensure crop security and the level of soil microbiological activity necessary for the health of the soil, however indicators like soil organic matter can be used for long-term measurements. To achieve sustainable agriculture development in India the following policy measures are suggested.

An analysis of area, production and productivity (yield) of agricultural crops (1974-2021) in India.

A number of crops are grown in the country depending upon the local agro climatic conditions. For simplicity of the analysis these crops have been divided into two broad categories such as Food crops, Nonfood crops.

This section deals with the trends in area, production of food crops, non-food crops, and Selected Agricultural crops in India during 1974 to 2021.

Area under agricultural crops

Agriculture can be defined as the systematic and controlled use of living organisms and the environment to improve the human condition. 'Agricultural land' is the land base upon which agriculture is practiced. Typically occurring on farms, agricultural activities are undertaken upon agricultural land to produce agricultural products. Although agricultural land is primarily required for the production of food for human and animal consumption, agricultural activities also include the growing of plants for fibre and fuels (including wood), and for other organically derived products (pharmaceuticals, etc.). Not all agricultural land is capable or suitable for producing all agricultural products, regardless of the level of management applied. The main limiting factors are climate and topography. Climate determines the heat energy and moisture inputs required for agricultural production.

Topographic limitations mostly restrict the ability to use cultivation equipment. Soils with all their variability are also a key limiting factor. Depending upon their properties and characteristics they may be appropriate for sustaining the production of certain agricultural products, but not others. This section intends to analyze the average change in area under agricultural crops in India.

A change in cropping pattern indicates a shift in area under the cultivation of agricultural crops. Land is one of the most important resources used in agriculture and continuous data for same is also available for a relatively longer period time. Resource diversification is discussed ,with the proportion of aggregate agricultural crops in the gross cropped area (GCA) of the country

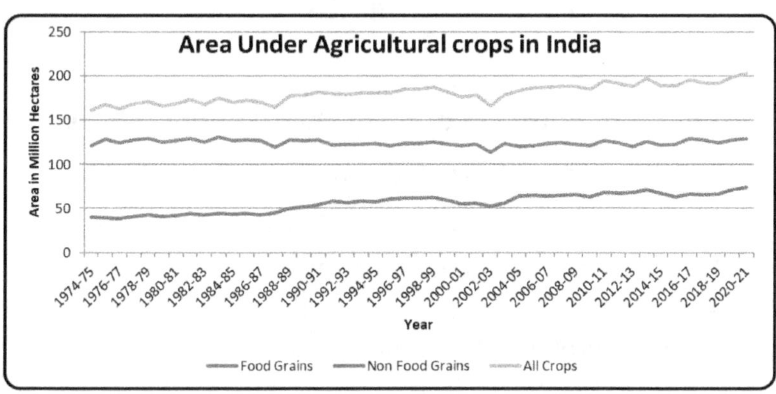

Figure:3.1: Area Under Agricultural Crops in India

Source: Hand book of Statistics on Indian Economy,RBI.

The figure shows that area under gross agricultural crops during the reference period. The area under food grains has stabled over a period of time. At the same time, the area under non- food grains have increased. So the area under all crops shows an upward trend due to increase in non- food grain crops.

Production of agricultural crops in India

Production means, the processes and methods employed to transform tangible inputs (raw materials) and intangible inputs (ideas, information, knowledge) into goods. Production is the output of a commodity created over a specified period of time.

In the earlier years of economic planning, food availability was a serious problem in India. Ninth five-year plan (1997-2002) emphasized on building of food stock to take up the challenge of famine and ever-increasing demand for food from the masses.

This section intends to analyze the average change in production of the agricultural crops in India.

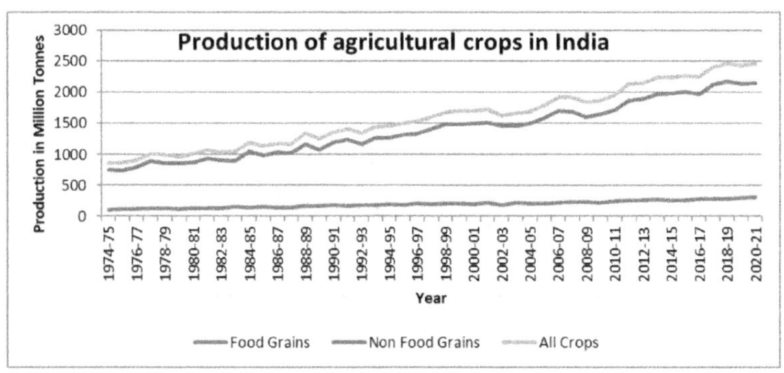

Figure:3.2: Production of Agricultural crops in India

Source: Hand book of Statistics on Indian Economy,RBI.

The above figure shows an increase in the production of gross agricultural crops over a period of time, except in case of some small fluctuations during the reference period. This ensures that sustainable Agricultural development in India.

An Analysis of Area under selected Agricultural crops in India

This section deals with the area, production and yield of selected agricultural crops, such as rice, wheat, cotton, sugarcane, potato and onion. The above stated crops are selected for the study on the basis of their significance in the total agricultural area and production, The study intends to analyze the average change in area, production and yield of the selected agricultural crops.

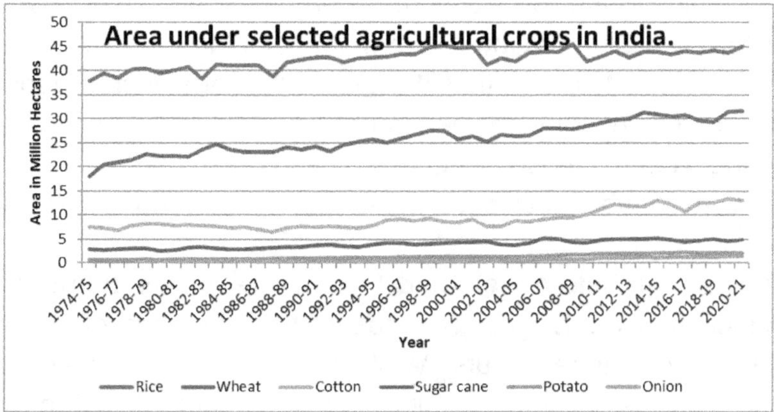

Figure: 3.3- Area Under selected Agricultural crops in India

Source: Directorate of Economics and Statistics, Department of Agriculture and Cooperation, Government of India. In the above diagram we notice that area under rice and wheat covers majority of the area under agricultural crops, the area under selected agricultural crops showing the fluctuating trend during the reference period due to natural, socio-economic, technical and institutional causes of cultivation of agriculture.

An Analysis of Production of select Agricultural crops.

The Present section deals with the average Production of selected Agricultural crops and discusses the extent of diversification of the production basket for individual crop.

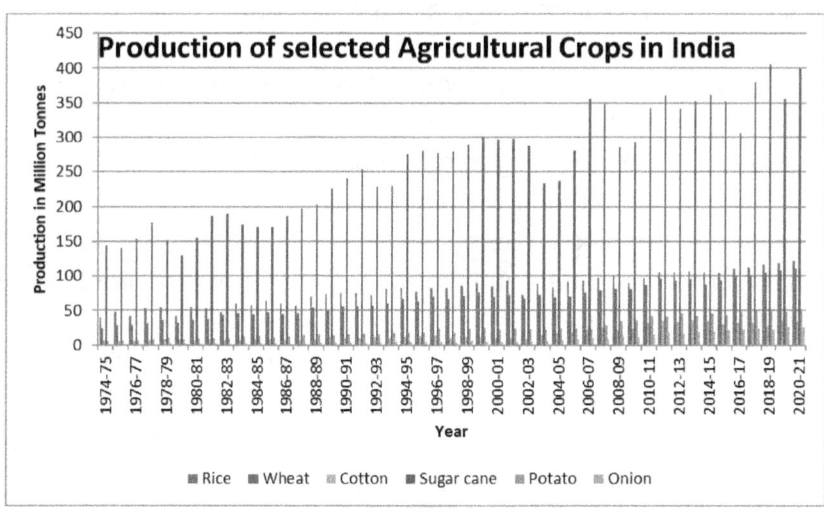

Source: Directorate of Economics and Statistics, Department of Agriculture and Cooperation, Government of India.

The figure 3.4 shows that the production of selected agricultural crops is increasing over the period of time, due to introduction of high yield variety seeds, modern technology,and other international factors. It shows sustainable development in Agriculture Production in India.

An Analysis of Yield of selected Agricultural crops in India:

Agricultural productivity is measured as the ratio of agricultural outputs to agricultural inputs. While individual products are usually measured by weight, their varying densities make measuring overall agricultural output difficult. Therefore, output is usually measured as the market value of the final output, which excludes intermediate products such as corn feed used in the meat industry. This output value may be compared with many different types of inputs such as labor and land (yield). These are called partial measures of productivity.

Agricultural productivity may also be measured by what is termed total factor productivity (TFP). This method of calculating agricultural productivity compares an index of agricultural inputs to an index of outputs. This measure of agricultural productivity was established to remedy for the partial measures of productivity; notably that it is often hard to identify the factors that cause them to change. Changes in TFP are usually attributed to technological improvements.

This section discusses with Average yield of selected Agricultural crops. The possibility of increasing agricultural yield can be understood under the following agrarian packages.
• Institutional change in the farming sector.
• Increase in the cropping intensity of land.
• Shift in the cropping pattern in favor of crops with higher productivity.
• Improvement in the technique of cultivation.
Moreover, with the application of technical innovation (like introduction of new HYV Seeds technology), institutional reforms of extension facilities (provision of irrigation facilities, better support services through government extension agencies); there can be a yield rate induced cropping pattern change.

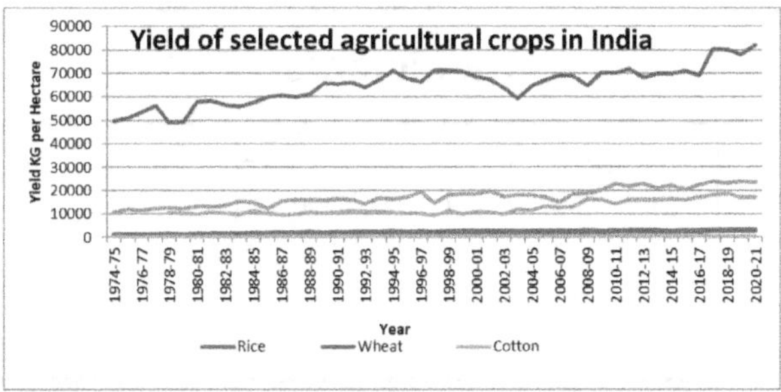

Figure:3.5 Yield of selected agricultural crops in India

Source: Directorate of Economics and Statistics, Department of Agriculture and Cooperation, Government of India.

The above figure shows that the yield of selected agricultural crops has been increasing during the reference period. This is because cropping is influenced by physical factors like irrigation, improved varieties, availability of fertilizers, influence of globalization and other institutional factors. It is observed that there is sustainability in Productivity of Indian Agriculture during the reference period.

Hypothesis testing:

To test the Hypothesis of how the priority in the utilization of agricultural land is being shifted from food to nonfood crops, we can formulate hypothesis below as follows:

1. Ho: There is no shift of agricultural land from food to non-food crops.
2. Ha: There is a shift of agricultural land from food to non-food crops.

Coefficients						
Model	Unstableized Coefficients		Standardized Coefficients	t	Sig. .263	R Square
	B	Std Error	Beta			
1. (Constant)	125.853	.612		205.521	.000	
Time period of Area under Food grains	-3.269	.866	-.513	-.3.775	.001	263
a. Dependent Variable: Area Under Food grains						

Table -4.1Result of dummy variable regression model for area under food grains

<table>
<tr><td colspan="8" align="center">Coefficients</td></tr>
<tr>
<td rowspan="2" colspan="2" align="center">Model</td>
<td colspan="2" align="center">Unstableized Coefficients</td>
<td align="center">Standardized a Coefficients</td>
<td rowspan="2" align="center">t</td>
<td rowspan="2" align="center">Sig.</td>
<td rowspan="2" align="center">R Square</td>
</tr>
<tr>
<td align="center">B</td>
<td align="center">Std Error</td>
<td align="center">Beta</td>
</tr>
<tr>
<td rowspan="2" align="center">1.</td>
<td align="center">(Constant)</td>
<td align="center">46.570</td>
<td align="center">1.277</td>
<td></td>
<td align="center">36.465</td>
<td align="center">.000</td>
<td></td>
</tr>
<tr>
<td align="center">Time period of Area under Non Food grains</td>
<td align="center">16.143</td>
<td align="center">1.806</td>
<td align="center">-.513</td>
<td align="center">.816</td>
<td align="center">.001</td>
<td align="center">.666</td>
</tr>
<tr><td colspan="8" align="center">a. Dependent Variable: Area Under Non-Food grains</td></tr>
</table>

To test the above hypothesis the study has employed the dummy variable regression model results. The dummy variable regression model results suggest that the computed value of nonfood grains is greater than the critical value and probability value is less than 1% level. So we did not accept the null hypothesis which suggests that there is no shift of land from food to non- food crops . On the contrary, we accept the alternative hypothesis which indicated that there is a shift of agricultural land from food to non- food crops.

This is because the farmers shift their land from growing food crops to commercial crops because the smaller the size of land holdings, higher the percentage of area under non-food grains. Moreover, the objective of income and profit maximization is foremost and the farmer is influenced in the choice of his crop by price and cost differences between different commodities. Other factors include the influence of globalization and other international factors.

The most important consideration affecting cropping pattern is the economic consideration even in a country like India which is dominated by farmers steeped in poverty and conservatism and where farmers hold fragmented bits of land, cropping pattern can be changed through appropriate change in economic motive. Experience in recent years has been that the farmer does accept the logic for a change wherever he is shown a better cropping pattern. The real difficulty in adopting a better cropping pattern is that the farmer may not have the requisite capital to invest in the present or possess the know-how that may be necessary for changing the crops. It is here that the government could come to his help.

Conclusions and policy imperatives:

With the given technological advances and capital investment in the farm sector, crop diversification can be an effective strategy to avoid early operation of the Law of Variable Proportions. If crops are diversified, increasing returns can be availed and the operation of the Law of returns can be postponed. Although Government has drafted various policies to this end viz. Soil Health Card scheme, Pradhan Mantri Fasal Bima Yojana, Dugdha Niti, Pashu Vikas Niti, etc. there are many areas which require urgent attention. Farm electrification for increased access to cheap irrigation facility, increased investment in fertilizers, seeds, agricultural machinery and infrastructural facilities, and sensitization of farmers through research and extension services are some of the areas which have huge potential to make diversification a success.

Summary

Cropping pattern and Crop diversification are intended to provide a wider choice in the production of a variety of crops in a given area so as to expand production related activities on various crops and also to lessen the risk. Crop diversification in India is generally viewed as a shift from traditionally grown, less remunerative crops to more remunerative crops.

The crop shift (diversification) also takes place due to governmental policies and thrust on some crops over a given time, for example: the creation of the Technology Mission on Oilseeds (TMO) to give thrust on oilseeds production as a national need.

This is done to facilitate lower national imports. Market infrastructural development and certain other price related supports also induce crop shift. Often ,low volume high-value crops like spices also aid in crop diversification.

Higher profitability and also the stability in production also induce crop diversification, for example sugar cane replacing rice and wheat. Crop diversification and growing multiple crops are practiced in rain fed lands to reduce the risk factor of crop failures resulting from drought or less rains. Crop substitution and shifts are practiced in places with distinct soil problems. For example, the growing of rice in high water table areas replacing oilseeds, pulses and cotton; promotion of soybean in place of sorghum in verticals (medium and deep black soils) etc.

The study has taken the area, production and yield of total food crops and nonfood crops at the National level. Besides, it also considers area, production and yield of selected agricultural crops. The study has selected two food crops such as rice and wheat, two commercial crops such as sugarcane and cotton, and two horticulture crops such as potato and onion at the National level.

The study points out that the change in cropping pattern cannot be analyzed in isolation from changes taking place in the farming system. These changes are determined by factors such as land ownership, access to resources, labor relations, livelihood strategies, farming practices, traditions and culture. The main causes of the changes may be grouped into (i) population growth and change in family structure, (ii) state intervention through land reforms, acquisition of land, deforestation, public distribution system etc. (iii) modernization and commercialization of agriculture, (iv) labor market conditions, and (v) price factors.

The impacts of changes are (i) economic (changes in production, farm income, employment, women's participation etc.);
ii) social and cultural (cultivator-labor relation, negative attitude to agriculture, loss of traditional skills, etc.) and (iii). Environmental

(loss of local varieties of seeds, breeds, and trees, receding water tables, decrease in biodiversity, etc.). Although it is required to evaluate the accuracy of estimation in detail, the method adopted in this study has a potential to accumulate the information on land use change, which is indispensable in environmental problems.

Findings

At the National level the area under food crops has decreased but area under non-food crops has increased so obviously it is noticed that area under all crops has increased.

It is observed that area under Rice, Wheat, Sugarcane, cotton; potato and onion have increased in the country.

The production and yield of food crops and non-food crops and selected agricultural crops has increased over the period of time. It shows that there is a sustainable development in the production and productivity of the agriculture crops.

Policy Measures for Sustainable Agricultural development.
- Declare a national policy for sustainable agriculture.
- Establish a national strategy for integrated pest management.
- Prioritize research into sustainable agriculture.
- Grant farmers appropriate property right.
- Direct subsidies and grants towards sustainable technologies.
- Link: support payments to resources conserving practices.
- Penalize polluters.
- Provide better information for the consumers and the public.
- Encourage the adoption of natural resource accounting.
- Establish appropriate standards and regulations for pesticides.
- Support farmers' training schools.
- Strengthen the capacities of NGOs to scale up.
- Rethink: the project culture.
- Scientific method of cultivation with best inputs including credit must be provided.

For achieving success in promoting sustainable development in agriculture, attention must be focused on land, water energy, nutrient supply, genetic diversity, pest management, systems approach and location of specific research and development.

Conclusion

India, being a vast country of continental dimensions, presents wide variations in agro-climatic conditions. Such variations have led to the evolution of regional niches for various crops. Historically, regions were often associated with the crops in which they specialize for various agronomic, climatic, hydro-geological, and even, historical reasons. But, in the aftermath of technological changes encompassing bio-chemical and irrigation technologies, the agronomic niches are undergoing significant changes. With the advent of irrigation and new farm technologies, the yield level of most crops, especially that of cereals, has witnessed an upward shift making it possible to obtain a given level of output with reduced area or more output with a given level of area and creating thereby the condition for inter-crop area shift (diversification). Without much disturbance in output level. Besides, as agriculture has slowly but surely become drought proof and the growth has become more regionally balanced, there has been a reduction in the instability with respect to agricultural output.

In the face of these new changes including the achievement of food self-sufficiency, the area shift that tended towards cereals in the immediate aftermath of the Green Revolution, has started moving in the opposite direction, i.e., from cereals to non-cereals. Although these reverse area shifts actually took place in the mid-1970's as a part of the process of commercialization, they became more pronounced since the mid 1980's as a response partly to emerging supply deficit in major crops and partly to the changing comparative advantage of crops. Since the recent trend in inter-crop area shifts has its origin in the price and trade policy changes of the 1980's, they indicate the increasing market influence on area allocation. The area under commercial crops has increased in the last three decades. Among the food grain crops, the area under superior cereals, i.e., rice wheat, is increasing; like any other economy, the share of agriculture in the GDP is also declining in India. Increase in income from the agriculture sector, further growth of non-crop sub-sectors within agriculture; faster growth of non-food grain crops; and faster growth of superior cereals. among the food grains are all happening, but the pace of such change is far too slow.

An accelerated pace of diversification to create positive import of higher income, higher employment and conservation and efficient use of natural resources emphasize the need for efficient policies, especially in technological development, selective economic reforms and institutional change. A strategy of crucial importance is growth enhancing non-farm activities. This calls for investment in rural infrastructure and skill up gradation and it also implies a careful examination and adjustment of macro-policies, which influence the relative profitability of different activities and in turn determine the nature and pace of diversification. In order to ensure social equity, policies on structural adjustment and reforms must pay special attention to the band of marginal and small farmers and agricultural laborers. The direct benefits from diversification should reach these sections of the farmers.

Bibliography: Books

1. AlaghY.K(2003): "Globalization and Agricultural crisis in India", Deep and Deep publication Pvt Ltd, New Delhi.
2. Bilgrami .S.A.R (1996): "Agricultural Economics", Himalaya Publishing House- Delhi and Nagpur.
3. Damodhar N. Gujarathi, Sangeetha (2008): "Basic econometrics", McGraw Hill Publishing Company Limited-New Delhi, 4th edition,
4. Desai. R.G (1998): "Agricultural Economics – models, problems and policy issues" Himalaya Publishing house, Bangalore.
5. Dhawan S.K (2006): "A handbook of Indian economy", Volume 1, Maxfored books publications, New Delhi.
6. Krishnaswamy O.R and M. Ranganathan (2010): "Methodology of research in social sciences", Himalaya Publishing house- Bangalore and Mumbai, Second revised edition (2005).
7. Mishra and Puri (2007): "Indian Economy", Himalaya Publishing House, 25th silver jubilee edition (2007).
8. Ranjan K.P (2000): "Agriculture growth and economic stability", Rawath Publications – Jaipur and New Delhi.
9. RuddarDatt, K.P.M Sundharam (2007): "Indian Economy", S. Chand and Company Ltd, fifth Edition.
10. Srinivasagowda M.V and Susheela (1996): "Studies in Agricultural Development", Deep and Deep publication Pvt Ltd, Bangalore and New Delhi.

Journals

1. Ashok Gulati, AniSharma(2006): "Freeing Trade in Agriculture Implications for Resource use Efficiency and cropping pattern changes" Economic and Political Weekly, A-155, December 27, 1997.

2. Balasubramaniana.M (1963): "Economics of cropping pattern" Indian journal of agricultural economics", Vol.36, No.1.

3. Bastime, c.c. and K.P Palnisami (1994): "An analysis of growth trends in principal crops in kerala" – Agricultural situation in India, Vol.48, No.12.

4. BrajeshJha, Nitesh Kumar and BiswajitMohanty(2009): "Pattern of Agricultural Diversification in India",Institute of Economic Growth, University of Delhi Enclave, North Campus Delhi – 110 007, India

5. BrijBala and S.D Sharma(2005): "Effect on Income and Employment of Diversification and Commercialization of Agriculture in Kullu District of Himachal Pradesh", Agricultural Economics Research Review,Volume 18, Number 2, July-December 2005.

6. Chellam Subramanian: "Sustainable Development Indian Agriculture" their Project Journal of Arts and Science. Nov, 2003. pp.27.

7. Desai B.M (1977): "Analysis of Cropping Pattern Farm Families Surat district" " Indian journal of agricultural economics" Vol.32, No.1.

8. E.utpalkumar (2003):"changing cropping pattern system in theory and practice: an economic insight into the agrarian west Bengal" " Indian journal of agricultural economics" Vol.36, No.1.

9. Joseph K. J. (1996): "Kerala's agriculture: its evolving structure with respect to cropping pattern changes-A markov Chain Analysis", Part of doctoral thesis in Agricultural Economics, Kochi.

Chapter 3

ISBN: 978-81-948672-7-2

Recent Floods and Associated Atmospheric and Oceanic Factors in Indian Region"
A Case Study in Climate Change Scenario

MANJUNATHA S TYALAGADI
Assistant Professor,
Dept of Geography,Shri C N Nirani
Government First Grade College,
Savalagi,Tal-Jamkhandi, District
Bagalkote,Karnataka

MAHANTESH RADDERATTI
Assistant Professor,
Dept of Geography,Shri C N Nirani
Government First Grade College,
Savalagi,Tal-Jamkhandi, District
Bagalkote,Karnataka

Email: *manjunathkiran2000@gmail.com; mantugeography@gmail.com*

Abstract: The Intergovernmental Panel for Climate Change has confirmed human induced warming will persists for centuries to millennia has continued to long term changes in the Climate Systems - Sea level rise, Floods, droughts and other aberrant conditions. These unusual and extremes are due to the interaction between atmospheric and oceanic factors. These complex weather systems influenced by both global atmospheric circulations and regional meteorological forces makes the monsoon an important piece of the climate puzzle. Any change in the system affects the food and water security of billions in the Indian subcontinent, many of them extremely poor.

The study aims in identifying the common features observed during the recent flood years in India. Datasets pertaining to Atmospheric and Oceanic conditions were obtained from NCEP Reanalysis. Mean Monthly, Anomalies, Long term average/climatology maps were prepared for the recent flood years. Majority of the flood years were associated with troughs/ cyclonic circulation sex tending up to 500 hPa as well as negative omega anomalies over the entire region indicating strong convectional currents at the surface. Analysis of SST revealed that majority of flood years were associated with positive anomalies over the three sub-regions and opposite conditions prevailed during droughts. High rainfall activity during the flood years was indicated by lower OLR values and vice-versa. It is observed that there are similarities in the flood years which will be of great help towards forecasting.

Keywords: Climate Change, Floods, Atmospheric, Oceanic, Anomaly maps.

Introduction

Extreme weather events like heat and cold waves, floods and droughts, sea-level rise, wildfires and vector-borne diseases affect the Earth's environment and its inhabitants adversely.Observational studies show an increasing trend in heavy to extremely heavy rainfall events over the Indian mainland in recent decades, particularly in the central Indian region (Roxy et al., 2017). On the contrary, Guhathakurta et al. (2015) have reported that the frequency of heavy rainfall events is decreasing in eastern central India and north India while increasing in peninsular, east, and northeast India.

The southwest monsoon (June to September), rainy season is known for its regional heterogeneity in terms of the rainfall distribution. Extreme precipitation events are on the rise in India which are driven by warming temperatures and thereby changes in the monsoon. The resulting floods are being exacerbated by unplanned urban growth and environmental degradation, driving millions from their homes and causing widespread damage.

For centuries, Indians rejoiced at the monsoon onset to break summer's high temperature. From June to September, rains water the crops, revive rivers and wells. Increasingly, however, the season's sweet relief is laced with apprehension. The torrential rains that submerged parts of India are the latest in a string of major floods. There are events leading to record breaking rainfall activity— a scenario that many worry could become the "new normal" as climate change increases the frequency of extreme weather. When monsoon arrives late, there is the march of a drought. Rains sputtered through June, then came on with a fierce intensity. July seems to receive spells of heavy rain leading to floods in many states, taking lives and displacing millions. Many farmers desperate for rain saw their crops washed away.

India's summer monsoon has always been variable and has often marked by floods- in the basins of the great Himalayan rivers. In fact, this anomalous monsoonal rainfall is a mix of global warming, unplanned urban growth and environmental degradation that leads to increasing flood risk in India. Recent studies show extreme precipitation events are on the rise in large parts of India, especially multi-day deluges that lead to large-scale floods. Warmer temperatures are speeding up glacier melt in the Himalayas, which is going to increase flow rates in the Ganges and Brahmaputra Rivers. Further, chopping of mountains and hills, as well as development on floodplains, are exacerbating risk, as was observed in the Kerala floods. These floods were caused by extreme rainfall and mismanagement of dam reservoirs, mining and construction and fragile mountain slopes. The floods in August 2018 took 483 lives, affected 5.4 million people.

Floods in India have increased from 67 (1996 to 2005) to 90 (2006 to 2015), according to the UN Office for Disaster Risk Reduction. India needs towards regulation of housing and infrastructure in floodplains — a trend that has to be intensified in recent years as India's urban population and economy have grown.Roxy Mathew Koll, states that in parts of India, there are more frequent heavy rainfall events as well as dry spells, and fewer moderate rainfall events.

Koll 2017 showed a three-fold rise in extreme precipitation events across central India from 1950 to 2015. He explained why extreme precipitation increased even as monsoon circulation weakened: The heavy rain events are caused by surges of moisture from a warming Arabian Sea.Most studies have confirmed that extreme rainfall will rise with temperatures. But, the models don't agree in terms of total rainfall. Roxy assures with high confidence that heavy rainfall events are going to increase.

Heavy rainfall over several days is what causes large-scale basin flooding as the rivers are fed by the Himalayas, glacier melt is an added worry.A study by analyzed satellite data on 650 glaciers across the Himalayas and found that the average ice loss rate had doubled during the 2000-2016 period, compared to the previous 25 years. More glacier melt means increased runoff, and more glacial lake outburst floods —

the sudden release of an enormous quantity of water when an ice block holding back water melts. In 2013, heavy rain followed by a glacial lake outburst caused devastating floods and landslides in the Himalayan state of Uttarakhand killing 4,000 people, damaged roads, bridges, and dams, causing economic losses of $3.8 billion.

On the India's coastal plains with deltaic plain with large urban populations — is another kind of flood risk could increase. Mishra et al found that short bursts of heavy rainfall, lasting only hours, are likely to increase by 20 percent if the global mean temperature rises above 1.5 degrees C. These extremes have the most impact on urban flooding and hence storm water systems need to be redesigned for the new extremes. Furthermore, Unplanned urbanization is already increasing flooding. Urban population has increased from 90 million people between 2001 and 2011 and further by 2050 it will increase to 416 million. Rain runoff automatically goes up when permeable soil is replaced by impervious surfaces - roads and parking lots. In addition, sprawling Indian cities have been building on wetlands and expanding into floodplains. And urbanization itself may be affecting rainfall patterns. Urban heat islands could be intensifying thunderstorms over Mumbai.In response to recent flooding events, the Maharashtra government has established a task force on climate change and Indian government is planning a new law to better manage major river basins.

Crying need includes modeling of small-scale cloud dynamics and river hydrology. Beyond forecasting, flood management policies need an overhaul. Flood management has been about creating embankments and dams to control floods. An audit conducted in 2017 across 17 Indian states shows delay in completion of river management projects and deficiencies in flood protection structures.

Meanwhile, new ecosystem-based flood management approaches being promoted by international agencies have to be implemented in India which includes planning cities with ponds and permeable surfaces, restoring wetlands and forests, and regulating development in floodplains and hills. Flood zone mapping and regulations would help the townships along Rivers, but have not been implemented.

Similarly, melting glaciers and shifting storm tracks played a major role in the disaster as global warming could lead to more catastrophic floods in the future. The early and heavy monsoon rains in the Himalayan region where houses are poorly built on the slopes, unregulated development along the Mandakini River that runs through Kedarnath, and soil erosion caused by thousands of pilgrims. There are so many illustrations to show. Keeping these aspects in view, an attempt is aimed to explain the possible association of upper air circulation, SST and OLR in causing flood over the country. This is achieved by the objectives as stated below-

1. Investigating upper air circulation in terms of convergence and divergence patterns
2. To assess the SST in relation with monsoon variability.
3. To determine the association of OLR during the abnormal monsoon years

This study will help to assess different patterns of the said parameters and their association with floods. This will enlighten to understand their influence on flood and the related abnormal atmospheric and oceanic parameters. These anomalous patterns would also serve as a signal for mitigating the problem of floods in future.

Data and methodology:

Meteorologists and Climatologists define flood (drought) in relation to the quantum of precipitation in excess (deficit) of that normally expected.India Meteorological Department (IMD) categories Flood or Drought years as the percentage departure from the normal rainfall.However, for the present analysis,this criteria is used to list out the Flood years during the period of 121years (1901-2020) (Parathasarthy,1984).

The formula for flood years is as below-

$$r>(\mu+\sigma)=Floodyear)$$

and drought years are when

$$r<(\mu-\sigma)=Drought\ year,$$

where, r=Rainfall,μ =Mean μ and σ=Standard deviation

Standard statistical methods such as mean,standard deviation and simple line arregression were used in order to detect increasing/decreasing trend in rainfall during the last 121 years. Moving average of 7 years was used to smoothen outshort term fluctuation and highlight long term trends/cycles.

The All India Rainfall for the period 1901 to 2016 was obtained IITM website and the extended data from 2017 to 2020 was downloaded from IMD website. The atmospheric and oceanic data considered for analysis are for the period 1970-2010. The circulation patterns at 500 hPa were obtained from http://210.212.167.213/chartsfor monthly and daily basis of the monsoon season during the flood years. These circulation patterns were studied and the locations of ridge/trough were identified and tabulated. Gridded (2.50 x 2.50) SST and OLR data over the Indian region (100S-400N & 500E-1000E) have been downloaded from http://esrl.noaa.gov/psd/during the flood years. The same website was used to download the Omega charts at 500 hPa in order to find upward and downward motions over the study region. Seasonal and monthly climatology maps were prepared for these parameters and their anomalies were worked out for each flood years. Further, daily SST and OLR anomaly maps were obtained and studied to know the duration of positive/negative anomalies.

Results and discussions
Identification of All India Flood years:
The rain fall data was scrutinized to identify the Flood years and the same is reported inTable1. It is clear from the table that there were in all 14 flood years during the last 121 years. The trend line runs parallel to the mean rainfall (Fig.1). Further, the rainfall data for India as a whole was subjected to seven and eleven years moving average and is illustrated in Fig.2.The left panel indicates increasing trend while right panel shows decreasing trend.

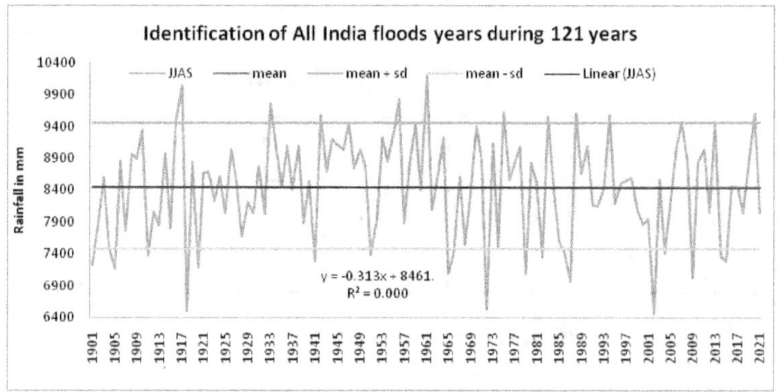

Fig.1 Identification of All India Flood Years

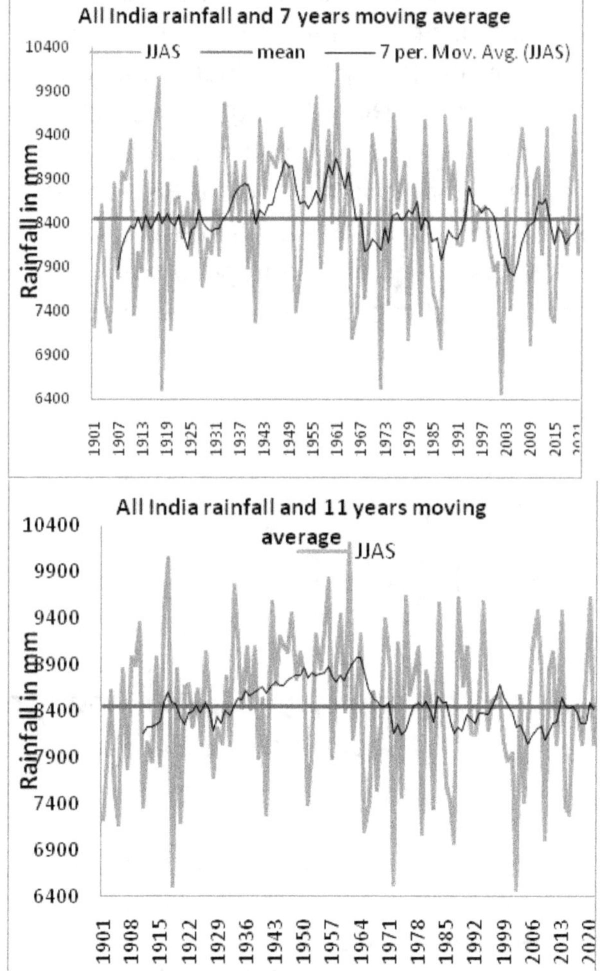

Fig.2 All India Rainfall and Moving Averages (Left Panel 7 yrs and Right Panel 11 yrs)

All India Flood years													
SI No	1	2	3	4	5	6	7	8	9	10	11	12	13	14
Flood Year	19 16	19 17	19 33	19 42	19 56	19 56	19 61	19 75	19 83	19 88	19 94	20 07	20 13	20 20

Table1. All India Flood years

Upper air circulations and other synoptic vortices:

In order to find out the circulation patterns, the presence of features like trough, ridge, anti-cyclone and cyclone over the Indian region were identified for All India flood years.Formation of systems (Cyclonic, Anti-cyclonic, Ridge/Trough as found in the flood year of 1988 are illustrated in Fig.3. The figure depicts the monthly anomalous circulation pattern for flood year of 1988, generally stronger than normal monsoon circulation is observed in June. In July, August and September, the lower and middle tropospheric circulation is indicative of weaker than normal circulation but upper tropospheric circulation is stronger than normal.

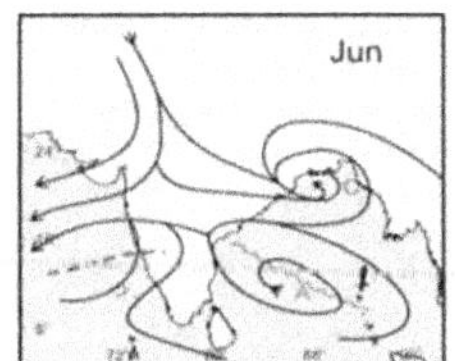

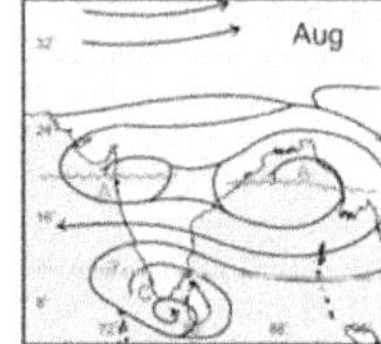

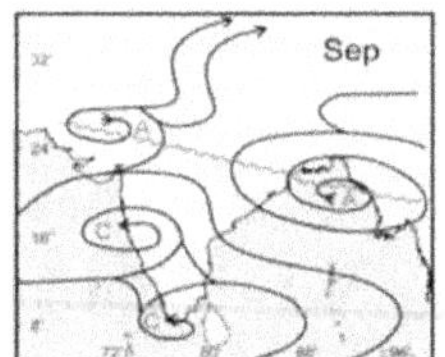

Fig.3 Monthly anomaly maps of Flood year(1988) observed at 500 hPa

The presence of troughs and ridges were studied for flood years and the results are given in table 2. The table suggests that majority of the flood years were associated with the troughs/ cyclonic circulations at all the levels, in general,and at 500hPa,in particular.

Floods year	June	July	Aug	Sept
1983	R(along 20-300N)	T(along 300-330N)	T(along300 N)	T(along 15-230N)
1988	0.057	0.18	T/C(along 100N), R/A(along 170N)	R/A (along 20-250N)
2007	T/C	T/C	R/A	T/C

Table 2 Formation of troughs(T)and ridges(R) over India during flood years
Note:(A-anticyclone and C-cyclone)

The above-said findings were further supported by analysis of Omega charts at 500 hPa level. Seasonal and monthly climatology maps for Omega (ω) component, along with their anomalies during each Flood years were analyzed. Fig.4 presents the seasonal climatology map and anomaly maps for a specific flood year (1970). It is clear from the seasonal climatology map that negative omega values prevail over most of the country, with its core lying over Bay of Bengal; while positive values are observed over NW region and adjoining areas. During the flood year (Fig. 4) negative anomalies are seen over the Indian Ocean, indicating divergence aloft while falling pressure and strong convectional currents at the surface In order to get more insights into the intra-seasonal variability of rainfall during each Flood year, monthly omega charts were studied.

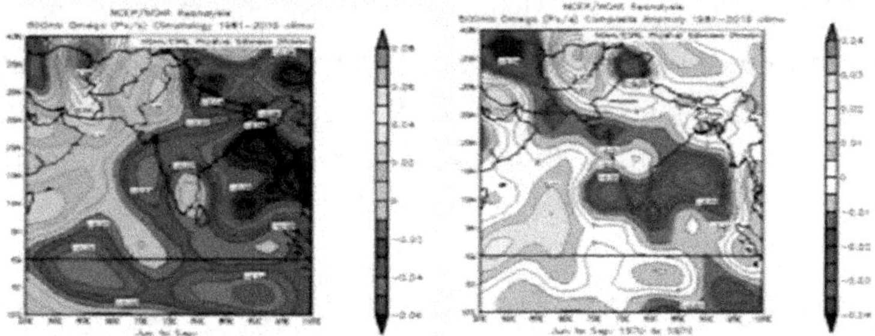

Fig.4 Seasonal climatology of Omega with anomalies during flood years 1988

Comparison of monthly anomaly maps for specific flood year (1970) and drought year (2002) with monthly climatology maps are depicted in Fig.5 (a to l) With reference to these figures; the following section discusses the month wise deviations of 1970 and 2002 from their respective monthly averages (normals). The climatology of June indicates that negative values are predominantly found over both the seas while positive values were observed overlying the heat low region. The negative values of Omega are more intensified over the Head bay region during the flood year. Conversely, positive anomalies prevailed over the same region during drought year (Fig.5a).

The monsoon sets over the country in the month of July and this is well- reflected in Fig.5d. In 1970, the month of July experiences negative anomalies over peninsular region and Bay of Bengal, but positive anomalies are seen over central and NW India indicating weaker monsoonal current over this region. This behavior also reveals the intra-seasonal and intra- regional variations of monsoonal current. While looking into the July anomaly of drought year of 2002, it was observed that the entire country was under the grip of subsiding air current. Similar patterns in the climatology map (Fig. 5g) were observed during the month of August. With respect to flood year, the negative anomalies were found over most of the country suggesting the intensification in the amount of rainfall. During the drought year of 2002, positive anomalies prevailed over peninsular region suggesting subdued rainfall activity; whereas negative anomalies were observed over West Coastal Region and Bay of Bengal.

The climatology of September month (Fig.5j) depicts the increase in positive anomalies over NW part of India, implying withdrawal of monsoon from that region. The monsoon remained active in this month during the flood year of 1970. On the contrary, positive anomalies existed over a larger area during 2002, which cumulatively contributed to produce phenomenal drought of this decade.

Sea surface temperature: Sea Surface Temperature (SST) is one of the key factors influencing the Indian monsoon system and its circulation.

The frequency of tropical cyclones is also dependent on the thermal, wind and moisture fields over the ocean. In this connection, Gray et al (1992) gave two important factors that influence the formation of Cyclones i.e., SST & gt;= 26 ° C and wind stress & gt;= 40 m 2 s -2 .

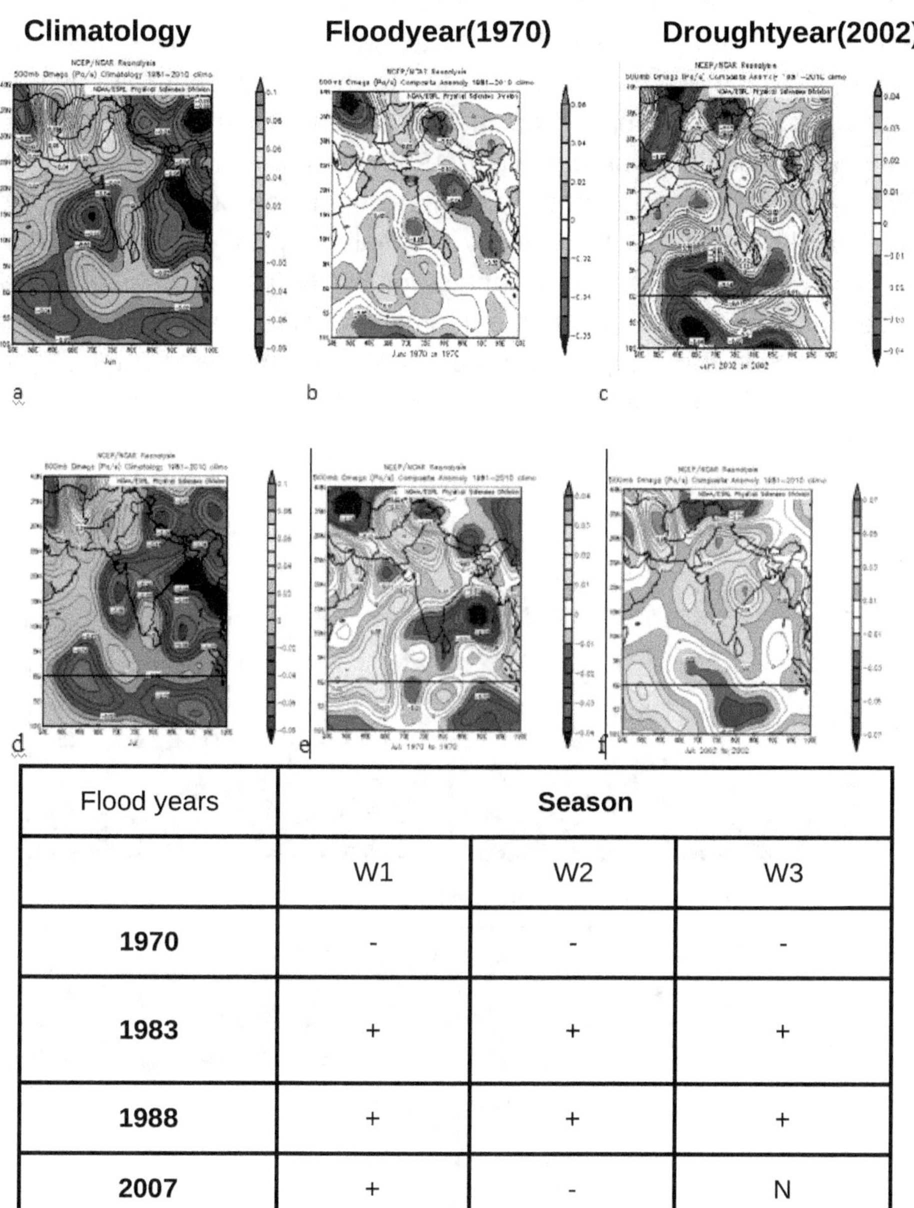

Climatology **Floodyear(1970)** **Droughtyear(2002)**

Flood years	Season		
	W1	W2	W3
1970	-	-	-
1983	+	+	+
1988	+	+	+
2007	+	-	N

Table 3 Seasonal SST anomalies during the flood years (Note: N= Normal)

The Table 3 depicts that majority of flood years (except 1970) are associated with positive anomalies over the three sub-regions, indicating strong convectional currents strengthening the monsoonal flow. The year 1970 exemplifies such a year, in which the regional intra- seasonal activity (formation of above normal LPS, no break days, early onset and late withdrawal) appears to have overcome the unfavorable large-scale factors related to the antecedent parameters. Monthly climatology maps and their anomalies for the same specific flood year (1988) and drought year (1986).are represented in the Fig.8. The same anomaly patterns which were observed on seasonal scale were well reflected in the monthly anomaly maps of 1988 and 1986. The same procedure was adopted for analyzing monthly anomaly maps for each flood year and the anomalies worked out are reported in the table 4. It is clear from the table that seasonal pattern for all the years, including the anomalous year of 1970, is replicated on monthly basis also.

a. Climatology

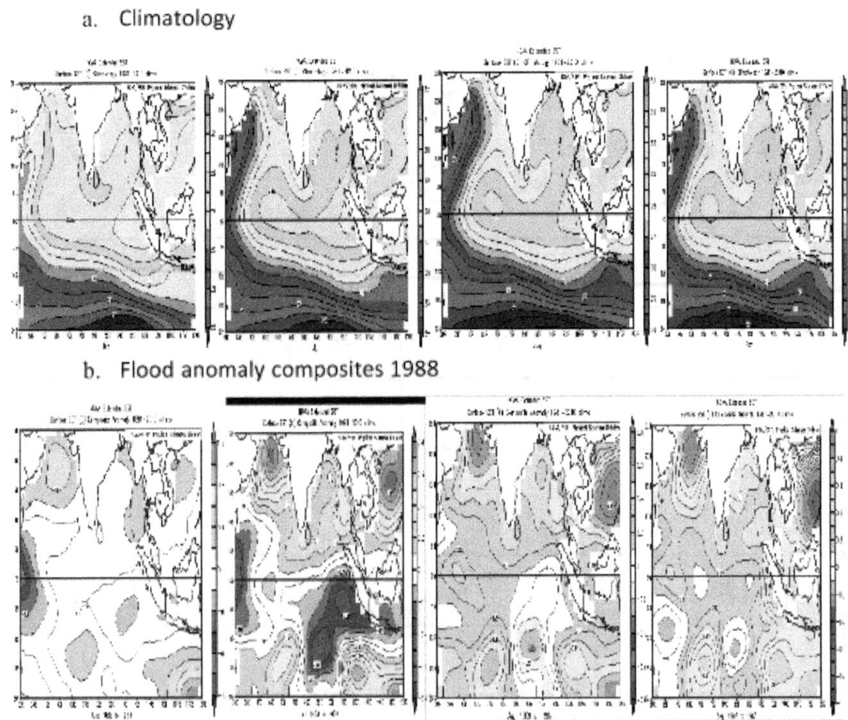

b. Flood anomaly composites 1988

c. Drought anomaly composites 1986

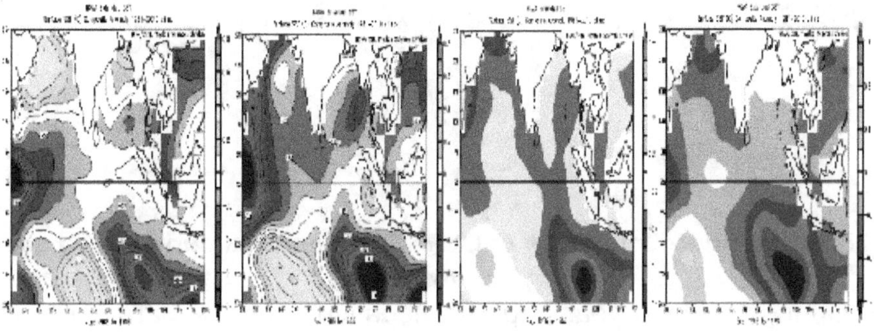

Fig. 8 a-c Monthly SST Climatology with anomalies of flood and drought year

Flood	June			July			Aug			Sept		
years	W1	W2	W3	W1	W2	W3	W1	W2	W3	W1	W2	W3
1970	-	-	-	-	N	-	-	N	N	-	-	N
1983	+	+	N	+	+	-	+	+	N	+	+	N
1988	N	N	+	+	+	+	+	+	+	+	+	+
2007	+	+	N	+	+	+	+	+	N	+	+	N

Table 4 Monthly SST anomalies during the flood years (Note: N= Normal)

Outgoing long wave radiation:

Outgoing long wave radiation (OLR) is an important factor to represent and understand cloudiness. Seasonal and monthly climatology maps for OLR, along with their anomalies during each F/D years were examined. Fig.8 depicts the seasonal climatology map and anomaly maps for a specific flood year (1970) and drought year (2002). It is clear from the seasonal climatology map that lower OLR values prevail over most of the country, with its core lying over Bay of Bengal; while higher values are observed over NW region and adjoining areas.

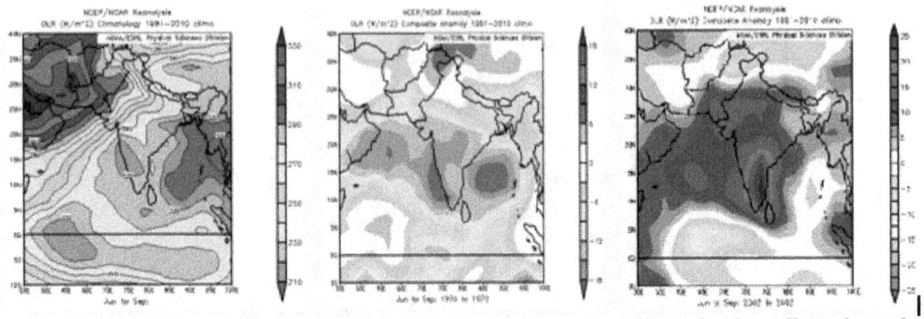

Fig.9 Seasonal climatology of OLR with anomalies during flood and drought years

During the flood year (Fig. 8b) negative anomalies are seen over the entire region, indicating thick cloudiness associated with the prevailing strong convectional currents. In order to get more insights into the intra-seasonal variability of rainfall during each F/D, monthly OLR charts were studied. Comparison of monthly anomaly maps for specific flood year (1970) and drought year (2002) with monthly climatology maps are depicted in Fig.10 (a to l).

A cursory glance over the figure implies a similar monthly pattern in relation with the seasonal maps. However, regional intricacies could be identified on monthly scale during the said F/D years. One such disparity can be located in Fig.10g, where unlike the monthly climatology; the flood year witnessed negative anomalies over NW region.

Similarly, during the month of September (Fig.10h) the monsoon was still active over the entire country, in general, and NW region, in particular. This can be attributed to the late withdrawal of the monsoon current from the region. This finding is supported by Mausam (1971), IMD publication. Another such regional variation can be found in the monthly anomaly map of August 2002, where negative anomalies were found over central India in spite of drought conditions prevailing over rest of the country. On similar lines, the anomalies for all flood years were compared with the monthly climatology map and tabulated in Table 5.

The table depicts that negative anomalies were observed during most of the flood years, suggesting less OLR values on account of the increase in cloudiness. However, positive anomalies are observed during some months in 1983 and 2007. In 1983, the onset of the monsoon was delayed and the monsoon current was weaker over NER and Saurashtra and Kutch region in the first fortnight of July. In 2007, anomalous southeasterlies were observed indicating subdued monsoon activity over the peninsular region.

Flood years	June	July	Aug	Sept
1970	India (-ve)	NW (+ve), Pen & NE (-ve)	India (-ve)	India (-ve)
1983	India (+ve)	NER&Guj (+ve), NW (-ve)	India (-ve)	India (-ve)
1988	NI&WCR (+ve), NE&Wcoast (-ve)	India (-ve)	India (-ve)	India (-ve)
2007	India(–ve)	NW& PEN (+ve), NE&Central India (-ve) (-ve)	PEN (+ve), WCR&NE (-ve)	India (normal and below)

Table 5: Monthly anomalies of OLR during flood years

Similarly, duration of daily anomalies of OLR during Flood years (June-July) were worked out for a specific Flood year (Fig.11).

The figure shows duration of daily positive or negative anomalies for all Flood years are tabulated in table 6. The results are in agreement with the patterns observed in seasonal as well as monthly anomaly maps.

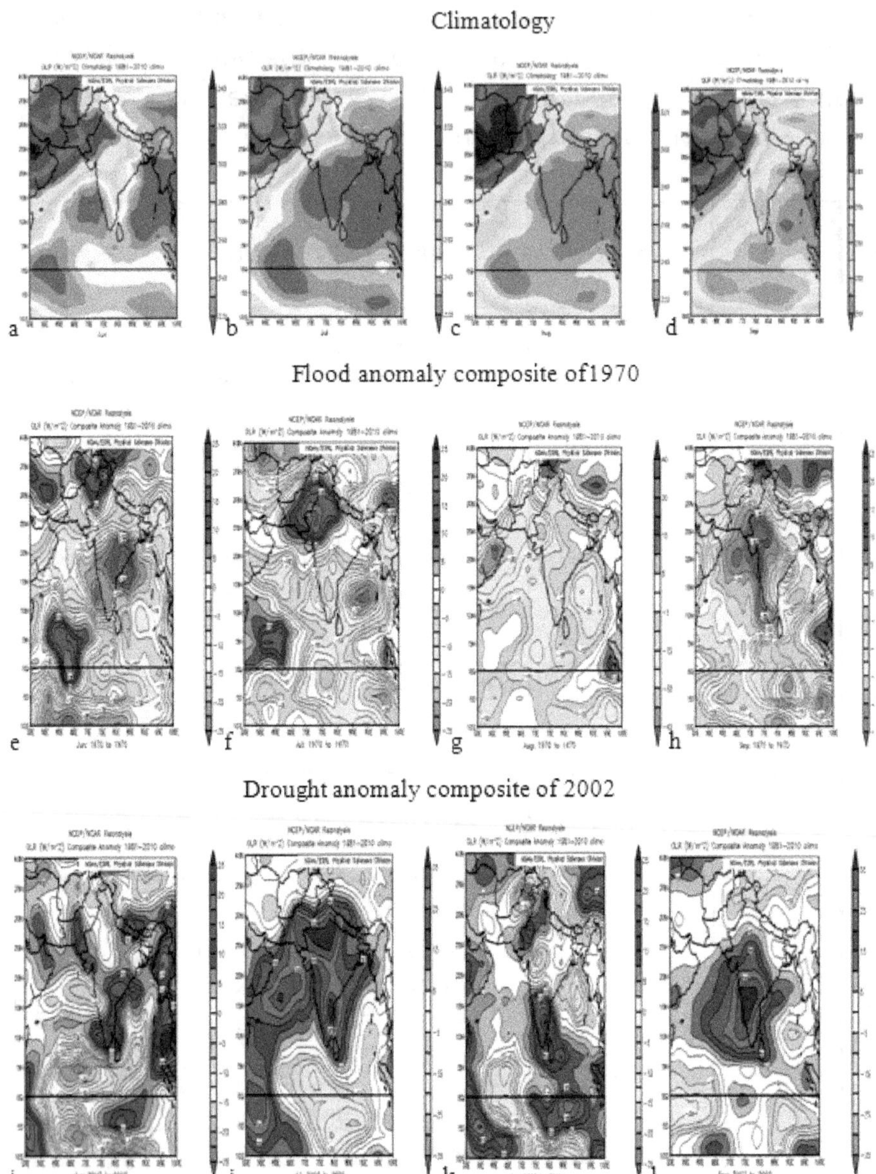

Climatology

Flood anomaly composite of 1970

Drought anomaly composite of 2002

Fig. 10 a-l. Monthly OLR climatology with anomalies of flood and drought years

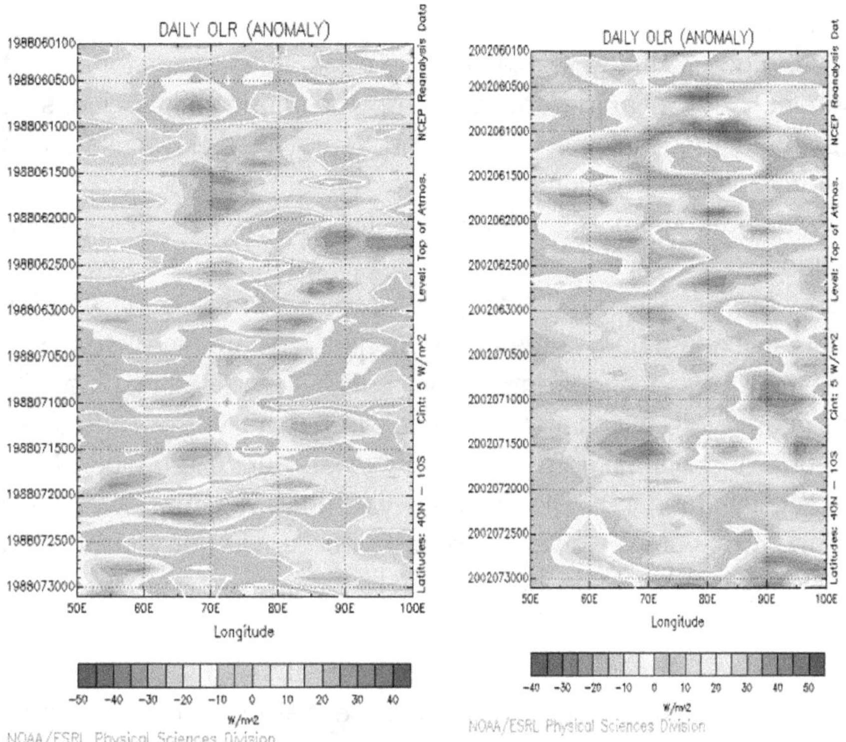

Fig.11 Daily OLR anomalies during 1970 and 2002

Flood year	Long.	Duration of Positive anomalies	No of Days
1983	60-90 degreeE	5/6-20/6,15/7-31/7	31
1988	60-90 degreeE	25/6-30/7	35
2007	60-90 degreeE	7/6-25/6	18

Table 6 Daily anomalies of OLR during Flood years (June-July)

Conclusion:

An attempt was made to examine the anomalous characteristics of atmospheric and oceanic parameters on the monsoon variations. It was observed that majority of the flood years were associated with troughs/ cyclonic circulations extending up to 500 hPa as well as negative omega anomalies over the entire region indicating strong convectional currents at the surface. On the contrary, the most of the droughts were associated with ridge or anticyclone, the occurrence of 18 trough/ cyclone during some of the months indicated intra-seasonal variations of monsoonal current. Analysis of SST revealed that majority of flood years were associated with positive anomalies over the three sub-regions and opposite conditions prevailed during droughts. High rainfall activity during the flood years was indicated by lower OLR values and vice-versa. However, there is no denying of the fact that regional intricacies could be identified on monthly scale.

Reference:

1. Roxy, M.K., Ghosh, S., Pathak, A. et al. A threefold rise in widespread extreme rain events over central India. Nat Commun, 8, 708 (2017). https://doi.org/10.1038/s41467-017-00744-9
2. Guhathakurta, P., Rajeevan, M., Sikka, D.R. and Tyagi, A. (2015), Observed changes in southwest monsoon rainfall over India during 1901–2011. Int. J. Climatol, 35: 1881-1898. https://doi.org/10.1002/joc.4095
3. Maurer, J. M., Sahaefer, J. M., Rupper, S., Corley, A., (2019), Acceleration of ice loss across the Himayalas over the past 40years, Science Advances, Vol.5(6), pp1-12.
4. Blenkinsop, S., Fowler, H. J., Barbero, R., Chan, S. C., Guerreiro, S. B., Kendon, E., Lenderink, G., Lewis, E., Li, X.-F., Westra, S., Alexander, L., Allan, R. P., Berg, P., Dunn, R. J. H., Ekström, M., Evans, J. P., Holland, G., Jones, R., Kjellström, E., Klein-Tank, A., Lettenmaier, D., Mishra, V., Prein, A. F., Sheffield, J., and Tye, M. R, (2018) The INTENSE project: using observations and models to understand the past, present and future of sub-daily rainfall extremes, Adv. Sci. Res., 15, 117–126, https://doi.org/10.5194/asr-15-117-2018.

- Parthasarathy, B.. (1984). Interannual and long-term variability of Indian summer monsoon rainfall. Proceedings of the Indian Academy of Sciences (Earth Planet Science). 93. 371-385. 10.1007/BF02843255.
- Gray, W. M., C. W. Landsea, P. W. Mielke, Jr., and K. J. Berry, 1992: Predicting Atlantic seasonal hurricane activity 6-11 months in advance. Wea. Forecasting, Vol.7, pp440-455
- Krishnakumar, G., 2006: Impacts of Indian Ocean temperatures on Indian monsoons with special reference to recent El-Ninos, PhD. Thesis Dept. of Environmental Science University of Pune , Pune.pp49.
- Parwaiz, Aslam., Parviainen, Janne., Somboon, Pannawadee., Mcdonald, Ariela, 2020, Disaster Rish Reduction in India, Status Report 2020, pp 10-18
- Report of the Conptroller and Auditor General of India on Schemes for Flood Control and Flood Forecasting, Ministry of Water Resources, River Development and Ganga Rejuvenation, Report No. 10 of 2017 pp9

Chapter 4 **ISBN: 978-81-948672-7-2**

Farmers Attitude Towards Sustainable Agricultural Practices

JAGADEESHA B

Assistant Professor
Department of Economics
Government First Grade College, Nanjangud,
Mysore, Karnataka
Email: jagabasava@gmail.com

Abstract: This study aims to identify the attitude of the future farmers regarding sustainable agriculture, based on the understanding of sustainable agriculture concept, its current and future importance and the fact that agronomy students will be future managers of rural enterprises, public managers and companies of the sector. Through information analysis obtained from the application of the attitude scale toward sustainable agriculture it was possible to notice that the aspects of this concept to the balance of environment and natural resources, and that through these practices agriculture can have both environmental and economic gains. Moreover, level of education, farm size, annual income, extension contact, and knowledge had positive and significant relationship with the attitude of the farmers towards sustainable agriculture. The results of the study showed that the higher the socioeconomic status (more frequent contact with extension services, higher education, ownership of land, etc.) and the greater the access to information, the greater the perceived importance of sustainable agricultural practices. It is concluded that if policymakers and extension organizations concentrate on these factors, they are more likely to succeed in making farmers more favourably disposed toward sustainable agriculture. It is expected that the next farmer's generation favourable attitude will generate greater implementation of the practices of sustainable agriculture in the long term.

Keywords: Agriculture, Environment, Farmers, Sustainable, Information, Welfare

Introduction

Sustainable farming is a broad, umbrella term for growing food using methods that will also nurture society, the environment, and the economy. It is an alternative to mainstream, industrial agriculture practices. Sustainable farmers seek to support community health and well-being and to work with nature, while still being profitable businesses—though farms can also be run as non-profits or recreational projects. Sustainable agriculture is using farming practices considering the ecological cycles. It is also sensitive towards the microorganisms and their equations with the environment at large. In simpler terms, sustainable farming is farming ecologically by promoting methods and practices that are economically viable, environmentally sound and protect public health. It does not only concentrate on the economic aspect of farming, but also on the use of non-renewable factors in the process thoughtfully and effectively.

This contributes to the growth of nutritious and healthy food as well as brings up the standard of living of the farmer. Our environment, and subsequently, our ecology has become an area of concern for us over the last few decades. This has increasingly led us to contemplate, innovate and employ alternate methods or smaller initiatives to save our ecology. Sustainable farming or Sustainable agriculture helps the farmers innovate and employ recycling methods, apart from the conventional perks of farming. A very good example of recycling in sustainable farming would be crop waste or animal manure. The same can be transformed into fertilizers that can help enrich the soil. Another method that can be employed is crop rotation. This helps the soil maintain its nutrients and keeps the soil rich and potent. Collection of rainwater via channeling and then its utilization for irrigation is also a good example of sustainable farming practices.

Objective of the Study

- To determine the selected socio-demographic characteristics of farmers.
- To determine Profit maximization on sustainable agriculture.
- To determine the attitude of farmers towards sustainable agriculture.

- To explore the relationship between selected characteristics of the farmers and their attitude towards sustainable agriculture.
- To determine the extent to which selected socio-economic characteristics influence the importance the farmers' attitude to sustainable agriculture.

Research Methodology

This chapter is basically descriptive and analytical in nature in this paper an attempt has been taken to analyse the attitude towards Sustainable Agriculture. The study is descriptive and analytical in nature. The study is basically based on primary and secondary sources, which include interview method, discussion, field study, questionnaire, books, journals and websites, relevant Articles, and magazines.

Sustainable agriculture

Thinking about technology, the first agricultural revolution was characterized by slow abandonment of fallowing and the introduction of crop rotation with legumes and/or tubers. These plants could be used both for soil fertilization, and in human and animal feeding. Thus, it was possible to intensify land use and achieve significant increases in agricultural production, "eliminating" chronic food shortages that characterized the earlier periods. The late nineteenth century and early twentieth century was also a period of intense changes in agriculture.

Great scientific discoveries, combined with the technological development of chemical fertilizers, internal combustion engines, plant breeding, among others, eventually imposed a new development pattern for agriculture.

This period is noteworthy with significant changes: the reduction of the relative importance of crop rotation; the progressive abandonment of green and animal manure use on soil fertility; the separation of vegetable and animal production and mainly the absorption of some stages of the agricultural production process by the industrial sector.

Thus, the opportunity for the development of more intensive production systems appeared, marking the beginning of a new stage in agriculture history. This new second stage is called contemporary agricultural revolution.

Starting from the 1st World War, the emerging chemical and mechanical industries intensified the production of inputs (pesticides, fertilizers, seeds, animal feed) and agricultural machinery (tractors, harvesters, plots), and agriculture started to depend less on local resources. The industrial sector then began to transform products from agriculture, industrializing them and distributing an increasing share of agricultural production.

Advances in transport processes, storage and agricultural product conservation enabled the emergence of a "unified" international market. Organic farming is also a type of sustainable farming. Because organic farming rejects the use of harsh pesticides and herbicides, and because of its focus on environmental health, it is generally considered a type of sustainable farming. The main components of both sustainable farming are exactly the same as soil Management, Crop Management, Water Management, Disease / Pest Management, Waste Management.

Significance of sustainable farming

Sustainable farming is important because it offers a solution to the problems caused by the way most of our food is grown today. Today's industrial farming methods, many stemming from the Green Revolution of the 1950s and 1960s, are depleting our natural resources through monocultures and the overuse of pesticides and fertilizers, among other practices, while leaving people with unequal access to food and nutrition around the world.

• Environment

Soil is considered a non-renewable resource, and sustainable farming promises to protect and preserve soil health.

• Public health

Putting food production in hands of disenfranchised communities, as sustainable farming advocates often do, is one way to correct food system injustices that result in continued health disparities among people of colour.

• Animal Welfare

Most animals raised for human consumption are grown and processed in conditions that are bad for their health. Sustainable farmers think about how to reform those industrial practices, such as reducing the use of antibiotics.

• Local economies and workers

Farmers and farmworkers are often exploited for their labour in industrial agriculture. The sustainable farming movement is creating the space for a food system that respects the dignity of farmers and workers.

Most efficient use of non-renewable resources

Coal, nuclear, oil, and natural gas are non-renewable energy resources that we use to drive cars and trucks, to cook, to heat our homes, and to run power plants that light our tablets and screens. The use of non-renewable energy resources is the leading cause of climate change, and food production is a primary sector contributing to greenhouse gas emissions. Sustainable farmers seek to be careful in their use of such resources, in alignment with their goal of protecting the environment.

Attitude

The idea that attitudes are dispositions to evaluate objects, people, or actions seem that an individual has one, and only one attitude toward any object or subject. In more recent works, it is suggested that attitudes can change, and when they change, a new approach can replace the previous one, but it does not necessarily exclude the existing one. According to this model of dual attitudes, people may simultaneously have different attitudes for a certain object in the same context, which may present an implicit or habitual attitude, and explicit.

Reinforcing the idea emphasize that attitudes are relatively consistent with the behaviour that they reflect. However, despite this consistency, attitudes are not necessarily permanent and can change over time. Thus, it is important to consider the influence of the situation in the attitudes and behaviour of the person. Situational influences are events or circumstances that, at a specific time, influence the relationship between attitude and behaviour. A specific situation can make people behave inconsistent with their attitudes.

The farmers had topmost attitude on the sustainable agriculture practices in respect of 'Application of cow dung and compost increase soil fertility' was the highest followed by 'Technology should be used as best as possible to increase efficiency of agricultural production'.

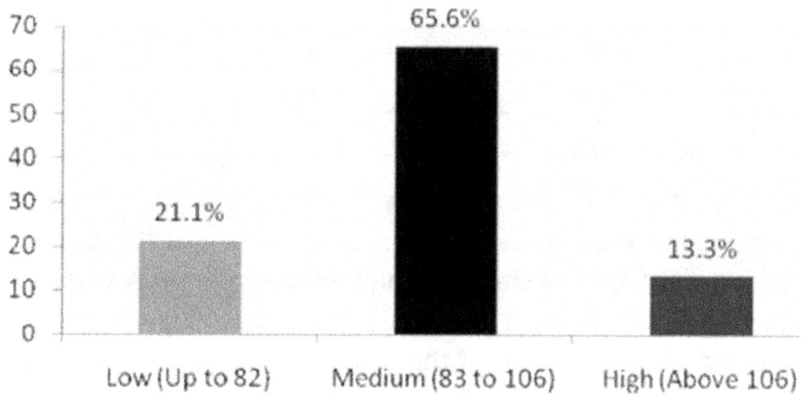

Figure 1: Distribution of the farmers according to their attitude Towards sustainable agricultural practices

From Figure 1 it is clear that majority (65.6%) of the respondents had medium attitude towards sustainable agriculture followed by 21.1% low and 13.3% high attitude respectively. However, it is observed that an overwhelming majority (79%) of the respondents in the study area had medium to high attitude towards sustainable agriculture. Most of the respondents of the study area knew the importance of sustainable agriculture. Many of them were agreed to the necessity of practicing sustainable agriculture to maintain a proper environment for agricultural production. They opined that, modern techniques like excess use of chemical fertilizer made the field unproductive and it is a threat to the agriculture.

Table 2. Relationship between selected characteristics of the farmers and their attitude towards sustainable agricultural practices

Selected personal attributes	Co-efficient of correlation (r)
Age	-0.137
Level of education	0.560**
Farm size	0.343**
Family size	0.111
Annual Income	0.293**
Farming experience	-0.190
Cosmopoliteness	0.549**
Extension contact Knowledge	0.400** 0.634**

Relationship between selected characteristics of the farmers and their attitude towards sustainable agricultural practices Coefficient of correlation was computed in order to explore the relationships between the selected characteristics of the farmers and their attitude towards sustainable agricultural practices. The null hypothesis was "there was no statistically significant relationship exists between the selected characteristics of the farmers and their attitude towards sustainable agricultural practices". Relationships between the selected characteristics of the farmers and their attitude towards sustainable agricultural practices have been presented in Table 2.

The question that had the highest rate of agreement and mean, that is, the one that the majority of students agreed with the statement, had the lowest standard deviation and variation coefficient. The question that addressed those sustainable agricultural practices (soil conservation, integrated pest management, decrease in the use of fertilizers and other chemicals) help to protect the environment and natural resources.

This showed that most respondents associated the concept of sustainable agriculture with aspects of environmental protection and natural resources, which is one of the pillars of this concept. Other statements that showed high average was related to use the non-renewable resources and other resources in the property in an efficient way, and when it is possible, integrate cycles and biological controls with the balance of the environment as a basis for sustainable agricultural practices. Besides that, farmers who practice sustainable agriculture live in greater harmony with nature. These statements follow the same analysis as before. The statements that obtained lower frequency said that the economic gains from the use of sustainable agricultural practices are not sufficient and that the farmer's net income may decrease when sustainable agricultural practices are implemented.

This indicates that respondents believe that sustainable agricultural practices can indeed generate greater income and economic gains to farmers, since the economic balance is inside this concept, that for economists, sustainable agriculture is synonymous of production maintenance and income of production physical systems, if possible, with low external inputs.

Conclusion

Can increased farmers attitude, knowledge and skills about sustainable agriculture with holding extensional workshop for farmers in area, create rural industries and industrial related to processing the agriculture production can decrease pressure on natural recourse and increase farmer's income in other hand. Access to more rural infrastructural and credits can increase job satisfaction between farmers. The majority of farmers were not participated in extension courses, and they do not access to agriculture credits.

Results of correlation test showed that older farmers with more age, experience in agriculture activities, family size and more agrarian land have low attitude toward sustainable agriculture than younger farmers.

In other hand off-farm income, farmers' knowledge about sustainable agriculture, level use of sustainable agriculture methods, extension contacts and job satisfaction have positive correlation with farmers' attitude toward sustainable agriculture. Most of the farmers of the study area were young aged, having large family size, small farm holding and had primary level of education with low to medium annual income and medium farming experience. Most of them had low extension contact. More than sixty percent (65.6%) of the respondents had medium attitude towards sustainable agricultural practices compared to 21.1% having low attitude and 13.3% had high attitude towards sustainable agricultural practices.

The findings revealed that farmers had a moderate attitude towards sustainable agricultural practices. This might be due to the fact that a considerable proportion of the farmers had not enough training exposure, moderate use of source of information and low extension contact. Coefficient of correlation test indicated that education, farm size, annual income, extension contact, and knowledge showed positive significant relationship with the attitude of the farmers towards sustainable agricultural practices that means higher the above-mentioned characteristics of the respondent, higher was their attitude regarding sustainable agriculture.

The objective of this research was achieved when it was possible to identify which factors are related to the future farmer's attitude to provide sustainable agriculture. The observed factors are related to viability, adoption and contribution of sustainable agricultural practices, environmental balance, and the paradigm of sustainable agricultural practices.

Through the analyses that were carried out, it was found that students identify the importance of sustainable agricultural practices and understand the concept, mainly related to the pillars of sustainable agriculture, the balance with the environment, rational use of non-renewable resources, and economic and social viability of these practices.

Through this research, it was possible to perceive that students associate the aspects of sustainable agriculture with environmental balance and natural resources, and that through these practices the farmer can have environmental and also economic benefits. Therefore, from this research it can be expected that the behaviour of the current agronomy students as farmers and future agribusiness professionals, may contribute to the sustainability of agriculture in the long term.

In terms of study limitations and recommendations for future research, it is suggested that this research be applied in an even larger sample, not only with students from agronomy courses, aiming to verify whether the factors found are the same, with the use of other scales to measure students' attitude toward sustainable agriculture. Finally, considering the importance of developing more sustainable behaviours aiming long-term agriculture and food production for the world population safely and in harmony with the environment, it is expected that undergraduate and technical courses in the agriculture field provide students with sufficient learning, so that they may adopt sustainable agricultural practices in the future.

References

Aktas, Y., 2001. Tarimsal yay2m surecinde tar2msal ilac saticilarinin yeri ve onemi: The place and importance of pest sellers in agricultural extension process.

Anonyms, 2009. Annual Meteorological Report. Iran Meteorological Organization, Khu zestan Meteorological Organization.

Duesterhaus, R. (1990). Sustainability's Promise. Journal of Soil and Water Conservation

Jabbour CJC (2010) Greening of business schools: a systemic view. International Journal of Sustainability in Higher Education,

Rasul, G. and G.B. Thapa, 2003. Sustainability of ecological and conventional agricultural systems in Bangladesh: An assessment based on environmental, economic and social perspectives.

Uddin, M. E. and M. M. Rahman. (2008). Students Perception Regarding Agro forestry and Global Warming.

Williams, D. L. (2000). Student's knowledge of and Expected Impact from Sustainable Agriculture.

Chapter 5   **ISBN: 978-81-948672-7-2**

Sustainable Food and Agriculture: Points to Ponder

DR. R.NAGABHUSHAN
Associate professor
Department of Economics, Government First Grade
College, Kuvempunagar, Mysore 570022.
Email: rnbhushan64@gmail.com

Abstract: Improvements in agricultural productivity have often come with social and environmental costs, including water scarcity, soil degradation, ecosystem stress, biodiversity loss, depleting fish stocks and forest cover, and high levels of greenhouse gas emissions. Today, 815 million people all over the world are hungry, and every third is malnourished. 2030 Agenda for Sustainable Development insists that that in order to overcome the complex challenges that the world faces require transformative action, embracing the principles of sustainability and tackling the root causes of poverty and hunger to leave no one behind. Food and agriculture form a vital connection between people and the planet, and thus can help achieve multiple Sustainable Development Goals (SDGs). The study suggests in order achieving sustainability in food and agriculture, a multi-dimensional approach is essential. Directing attention to strengthening the livelihoods of the poorest, building better rural-urban linkages and empowering rural people to become critical agents of change can lay the foundation to " leaving no one behind" The approach should be one of inclusiveness and far reaching.

Keywords: Agriculture, Environment, Productivity, Poverty, Sustainable, Development.

Introduction

Nothing, after national defence, is as important to the welfare of the world`s peoples as fostering agriculture. It is a matter of grave concern that food and agriculture stand today at crossroads. Improvements in agricultural productivity have often come with social and environmental costs, including water scarcity, soil degradation, ecosystem stress, biodiversity loss, depleting fish stocks and forest cover, and high levels of greenhouse gas emissions.

Today, 815 million people all over the world are hungry, and every third is malnourished. 2030 Agenda for Sustainable Development insists that that in order to overcome the complex challenges that the world faces require transformative action, embracing the principles of sustainability and tackling the root causes of poverty and hunger to leave no one behind. Food and agriculture form a vital connection between people and the planet, and thus can help achieve multiple Sustainable Development Goals (SDGs). By nurturing land and adopting sustainable agriculture, present and future generations will be able to feed a growing population. Agriculture, covering crops, livestock, aquaculture, fisheries and forests, is the world's biggest employer, largest economic sector for many countries, while providing the main source of food and income for the extreme poor. Sustainable food and agriculture have great potential to revitalize the rural landscape, deliver inclusive growth to countries and drive positive change.

Five Key Principles of Sustainable Food and Agricultural System:

Sustainable agricultural development will be realized only when food is nutritious and accessible for everyone, and at the same time it is ensured that natural resources are managed in a way that maintains ecosystem functions to support current and future human needs. Food and Agricultural Organisation (FAO) puts forward an approach that is based on five principles that balance the social, economic and environmental dimensions of sustainability, and provides a basis for developing adapted policies, strategies, regulations and incentives.

1. Increase productivity, employment and value addition in food systems.
2. Protect and enhance natural resources.
3. Improve livelihoods and foster inclusive economic growth.
4. Enhance the resilience of people, communities and ecosystems.
5. Adapt governance to new challenges.

FAO outlines 20 interconnected actions which are aligned to the five principles of sustainable food and agriculture. Each of these actions describe approaches, practices, policies and tools that interlink multiple Sustainable Development Goals (SDGs). These actions integrate the three dimensions of sustainable development – economic growth, social inclusion and environmental protection – and involve participation and partnerships among different actors. The 20 actions are integrated and interconnected. They knit together the many dimensions of agriculture and rural development with a country's broader development programme, laying the foundation for resilient and sustainable societies.

Key takeaways from the actions

Following is some of the key takeaways from the actions:

- **Improving productivity:** Improving agricultural productivity is key to transforming the livelihoods of hundreds of millions of people across the globe. Creating the conditions for inclusive rural transformation requires investing in basic infrastructure: roads, markets, land and water transportation, telecommunications and storage capacity. It means providing greater access to land, resources, services, finance, technologies and modern tools to generate energy.

- **Connecting smallholders to markets:** A fundamental part of any strategy towards more productive and sustainable agriculture and rural development is access for agricultural and food producers to markets with higher efficiency, transparency and competitiveness. More connected markets offer enormous opportunities to generate greater income.

- **Encouraging diversification of production and income:** Cultivating multiple crops brings significant benefits to society that go far beyond the farm gate. Not only does diversifying agricultural production conserve biodiversity, improve soil and plant health and reduce exposure to pests, diseases or extreme-weather events, it also brings greater rewards to farmers and the local community through improved nutrition, job creation and income generation.

- **Building producers' knowledge and developing their capacities:** Sharing knowledge, building capacities and investing in innovative technology are all part of the transformation to sustainable food and agriculture systems.

- **Enhancing soil health and restoring land:** Healthy soil produces healthy food and better nutrition. Sustainably managing soil is cheaper than rehabilitating or restoring soil functions.

- **Mainstreaming biodiversity conservation and protecting ecosystem function:** Biodiversity is integral to ecosystem health, important to increase food production and necessary to sustain livelihoods. Conserving and using a wide range of domestic plant and animal diversity provides adaptability and resilience in the face of climate change, emerging diseases, pressures on feed and water supplies and shifting market demands. Tapping into ecosystem services reduces the need for external inputs and improves efficiency.

- **Reducing losses, encouraging reuse and recycle, and promoting sustainable consumption:** Every year the world loses, or wastes, about a third of the food it produces. The global economic cost of food wastage is USD750 billion. But the impact is far greater. Food losses and waste affect both consumers and producers by raising the price of food and decreasing the amount that can be sold. They constitute a threat to food security, a waste of resources, ever-greater stress on ecosystems and a danger to the environment in the form of greenhouse gas-producing emissions. All actors in the food chain, from farm to fork, can play a role in reducing losses, reusing, recycling and promoting more sustainable consumption patterns.

- **Using social protection tools to enhance productivity and income**: Social protection plays a key role in reducing rural poverty and hunger. Measures such as cash and asset transfers provide liquidity and financial security to all poor people, giving them the means to invest in their future.

By providing a basic income, these measures help to relax insurance and credit constraints, allowing poor people the chance to start their own businesses, engage in profitable activities, and ultimately to break out of the cycle of multigenerational poverty.

- **Improving nutrition and promoting balanced diets:** Malnutrition in all of its forms – undernutrition, micronutrient deficiencies, obesity and diet-related non-communicable diseases – imposes unacceptably high economic and social costs on all countries. Creating nutrition-sensitive agriculture and food systems requires taking action at all stages of the food chain in order to deliver safe and nutritious food all year round to the consumer.

- **Addressing and adapting to climate change:** It is increasingly clear that the goals of achieving food security and sustainable agriculture and addressing the challenges of climate change are intertwined and need to be addressed in a coordinated manner. The capacity of the agriculture sectors to respond to climate change has far-reaching impacts on the livelihoods of the majority of people in many developing countries and on national economies.

Conclusion

The world, no doubt, has already been facing acute food shortage. The situation is sure to get worsen in future. In order to achieve sustainability in food and agriculture, a multi-dimensional approach is essential. Directing attention to strengthening the livelihoods of the poorest, building better rural-urban linkages and empowering rural people to become critical agents of change can lay the foundation to 'leaving no one behind'. The approach should be one of inclusiveness and far reaching.

Bibliography

- Transforming Food and Agriculture to Achieve the SDGs. Food and Agriculture Organization of the United Nations, 2018.

- International Guidebook of Environmental Finance Tools: A Sectoral Approach. United Nations Development Programme, 2012.

Chapter 6

ISBN: 978-81-948672-7-2

Sustainable Agricultural Systems and Practices in India

PREMAKUMARI. L

Assistant Professor of Economics
Government First Grade College
T.Narasipura-571 124
Email: premtmk@gmail.com

Abstract: Sustainable agriculture relies on replenishing the soil while reducing the use of non-renewable resources like natural gas which is used to convert atmospheric nitrogen into synthetic fertiliser and mineral ores. Measures to develop agriculture on a long-term basis can be taken on two fronts: soil and water. Soil maintains biological diversity and productivity, immobilises and detoxifies organic and inorganic materials, stores and cycles nutrients, and supports socioeconomic structure, among other functions. While sufficient rainfall is available in some areas for crop growth, irrigation is required in many others. To be sustainable, irrigation systems must be well-managed and must not use more water from their source than is replenished naturally; otherwise, the water source becomes a non-renewable resource. Improvements in water well drilling technology and submersible pumps, combined with the development of drip irrigation and low-pressure pivots, have made it possible to achieve high crop yields on a regular basis where relying solely on rainfall previously made this level of success impossible. The study suggests that Improvements in water well drilling technology and submersible pumps, combined with the development of drip irrigation and low-pressure pivots, have made it possible to consistently achieve high crop yields where previously relying on rainfall alone made this level of success unpredictable.

Keywords: Agriculture, Environment, Irrigation, Resources, Sustainable Development

Introduction

Agriculture has long been an important part of human development and the development of the nation as a whole. This primary sector providing basic needs for both human community and industries. The nature of agriculture nature has changed, and the priority has given to sensitize and adaptable to the local environment and ecosystem, but the role of agriculture in all places has become unified and equally important. The current state of agriculture has become increasingly important in light of the challenges that humanity and all living beings face. By 2050, the world's population will have surpassed 9.7 billion people, and the demand for food and fiber could be met by doubling or quadrupling the productivity of agricultural sector. In this context, after 1978, the sustainable development approach drew more attention, and after the 1992 Rio Earth Summit, it drew a larger audience. The Millennium Development Goals (MDGs) of 2000 also shed light on issues related to sustainable development, particularly the eradication of poverty and hunger.

In India, agriculture provides a living for more than 60% of the population. India has a total area of 328.7 million hectares, of which 58 percent is cultivable land. As per the Land Use Statistics 2016-17, there are 140 million hectares of net sown land and 192.2 million hectares of gross cropped land. The net irrigated area of the countryis 68.6 million hectares. Despite its low contribution to the nation's GDP, agriculture is the backbone of the Indian economy due to its high share of employment and livelihood creation. The share of agriculture to the gross domestic product has steadily decreased from 36.4 percent in 1982-83 to 19.9 percent in 2020-21. (Economic Survey 2020 - 21). Despite this, the sector continues to employ over half a billion people, accounting for 54.6 percent of the workforce (Census 2011). From 50.82 million tonnes in 1950-51, food grain production increased to 296.65 million tonnes in 2019-20. (4th Advance Estimates for 2019-20). It is also a significant source of raw materials and demand for a wide range of industrial products, including fertilisers, pesticides, agricultural implements, and consumer goods.

Sustainable Development

Since the early 1980s, researchers, environmentalists, and policymakers have been concerned about the issue of sustainable development. Its definition and goals have changed numerous times throughout the process. However, the World Commission on Environment and Development's definition has been the most popular so far (WCED). "Sustainable development is development that meets current needs without jeopardising future generations' ability to meet their own needs"(WCED, 1987). The concept of sustainable development has two dimensions: to improve (i.e. development) and to maintain (i.e.sustainability), and the primary focus of sustainability is on the issue of intergenerational equity, which implies that future generations have equal (or greater) options in terms of human well-being or production prospects than current generations.

For natural resource management, sustainable development is a multidimensional concept with three interacting angles: ecological security, economic efficiency, and social equity. Sustainable development entails not only the long-term viability of the environment and resource systems, but also the long-term viability of the economic and social systems. All development activities, regardless of scale, magnitude, or nature, have short- and long-term environmental consequences. Although the short-term effects may be insignificant in comparison to the benefits of developmental activities, the long-term effects must be considered because they may cause serious ecological and environmental problems, making long-term development impossible.

Sustainability in Agricultural Development

The word "sustain," derived from Latin word 'sustinere' (sus-from below and tenereto hold, to keep in existence or maintain), connotes long-term support or stability. Sustainable agriculture refers to farming systems that are "capable of indefinitely maintaining their productivity and usefulness to society." Socially supportive, commercially competitive, and environmental friendly and resource-conserving are some of the features of sustainable agricultural development.

Agricultural practises, like all developmental activities, have an impact on the environment. Agriculture not only has a significant impact on the environment, but it is also directly impacted by environmental changes (Tilman et al,2002). Aside from the obvious environmental issues associated with farming, the agricultural sector is the most important sector in developing economies. In these economies, this sector employs around 70% of the labour force and contributes about 34% of GDP on average. The most pressing issue in agriculture today is resource sustainability, as well as the indiscriminate use of chemical fertilisers and pesticides. As a result of these issues, there is a growing awareness of the need to move away from the input-intensive agriculture practised during the green revolution phase and toward sustainable farming in various parts of the world (Gautam and Bhardwaj, 2011). The widespread concern about the degradation and depletion of the natural resource base as a result of agricultural growth has led many to question its long-term viability.

The concept of linking agriculture and the environment will be addressed through meaningful farm research practises (Kuriakose&Iyer, 2011). Therefore, in a broader perspective, the term sustainable agriculture refers to an integrated system of plant and animal production practises with a site-specific application that, over time, will: meet human food and fibre needs; improve environmental quality and the natural resource base upon which the agricultural economy depends; make the most efficient use of non-renewable resources and on-farm resources; maintain farm operations' economic viability; and improve the quality of life for far-flung communities.

Measures to Achieve Sustainability in Agricultural Development

Sustainable agriculture relies on replenishing the soil while reducing the use of non-renewable resources like natural gas (which is used to convert atmospheric nitrogen into synthetic fertiliser) and mineral ores (e.g., phosphate). Measures to develop agriculture on a long-term basis can be taken on two fronts: soil and water.

Soil maintains biological diversity and productivity, immobilises and detoxifies organic and inorganic materials, stores and cycles nutrients, and supports socioeconomic structure, among other functions (Mandal & Sarkar, 2011). While sufficient rainfall is available in some areas for crop growth, irrigation is required in many others. To be sustainable, irrigation systems must be well-managed (to avoid salinization) and must not use more water from their source than is replenished naturally; otherwise, the water source becomes a non-renewable resource. Improvements in water well drilling technology and submersible pumps, combined with the development of drip irrigation and low-pressure pivots, have made it possible to achieve high crop yields on a regular basis where relying solely on rainfall previously made this level of success impossible.

Sustainability in Indian Agriculture

Despite the recent surge in manufacturing and services and the declining share of agriculture in national income, India can still be classified as an agricultural country because the majority of its workforce (approximately 65 percent) is still employed in agriculture and related activities. Since time immemorial, it has been India's noblest profession, and it has been practised on a long-term basis.

Large-scale forest areas, grazing lands, and waste lands have only recently been converted to croplands to support the rising population, resulting in ecological imbalance and air pollution.With no more agricultural land available, efforts have been made to increase food grain production using high-yielding varieties of seeds, fertilisers, and irrigation, as well as advanced farm equipment, dubbed the "green revolution" in India.

Aside from that, the so-called green revolution is limited to a few crops, namely wheat, rice, and maize, and has only been possible in a few areas, namely Punjab, Haryana, and Western Uttar Pradesh, as well as a few districts in Andhra Pradesh, Maharashtra, and Tamil Nadu.

While sufficient rainfall is available in some areas for crop growth, irrigation is required in many others. To be sustainable, irrigation systems must be well-managed (to avoid salinization) and must not use more water from their source than is replenished naturally; otherwise, the water source becomes a non-renewable resource. Improvements in water well drilling technology and submersible pumps, combined with the development of drip irrigation and low-pressure pivots, have made it possible to consistently achieve high crop yields where previously relying on rainfall alone made this level of success unpredictable.

Most green revolution regions have reached a productivity plateau, and farming profitability has begun to decline, though these regions remain highly productive in comparison to other regions and are critical to meeting future food demands (Vyas and Reddy, 1993). Furthermore, it has been observed that the high productivities achieved in green revolution regions are unstable and fluctuating (Mahendradev, 1987; Mitra,1990).

Furthermore, the failure to recognise the link between poverty and environmental sustainability has exacerbated the issue. Indeed, it is argued that well-designed poverty alleviation programmes could be a step toward a more environmentally friendly world (Vyas, 1991). As a result, the issue of agricultural sustainability in India can be examined from three perspectives: ecological, economic, and social.

Ecological Sustainability: Many traditional farm practises, as well as the majority of conventional farm practises, are not environmentally sustainable. They overuse natural resources, resulting in decreased soil fertility, soil erosion, and global climate change. Soil fertility, water, biodiversity, pollution, landscape, climate, and other factors all benefit from sustainable agriculture over traditional and conventional methods.

Economic Sustainability: Agriculture cannot be long-term sustainable unless it is economically viable. In terms of export vs. inward looking policy, debt, risk, employment, and so on, conventional agriculture poses greater long-term economic risks than "sustainable" alternatives.

Social Sustainability: Political and social unrest, inclusiveness, food security, gender parity in labour force participation, and other concepts of social acceptability and justice are all linked to the social sustainability of farming techniques.

Review of Literature

Reviewing literature on various aspects of agricultural sustainability, such as ecological, economic, and social sustainability, is necessary to build a foundation of knowledge that will allow the study to fill in the gaps that were discovered during the research study's exploration. In his article 'Food Security and the North-East,' Monirul Hussain (2004) assesses the food security situation in India's North-East region. He noted that the entire North-East has a food production deficit.

The Brahamputra valley, Assam's Bark valley, and Manipur's small Imphal valley are all densely populated, and the land-to-man ratio is becoming increasingly unfavourable.The concentration of land in the hands of a few has resulted in a significant increase in landlessness among the peasantry. As a result, a large number of landless peasants have lost their jobs and have no other means of surviving. Apart from that, a large segment of the population has lost their land and livelihood as a result of environmental and conflict-related displacement – the internally displaced persons (IDPs). In the North–East, IDPs are the most vulnerable people to food insecurity.

In their article 'Factors in Declining Crop Diversification – Case Study of Punjab,' J. Singh and R. S. Sidhu (2004) looked at the growth of agricultural output in Punjab and the role of crop shift and crop diversification in that growth. They calculated the diversification index for different regions of Punjab over time to show the scale of diversification. It was discovered that the state's overall diversification index fell from 0.707 in 1970-71 to 0.591 in 2001-02.

Rice and wheat continued to expand in area and production at the expense of other crops, and the output mix's diversity decreased over time.

As a result, the wheat–rice system became almost specialised throughout the state, with the degree of specialisation varying marginally across regions due to land and water constraints. With the current crop pattern and technology, agriculture's future growth will be primarily driven by area expansion, which is constrained by water scarcity. Otherwise, the crop pattern must be changed to include high-value crops such as fruits and vegetables.

Pant (2005) notes in his article 'Control of and Access to Ground Water in UP' that the number of private tube wells (PTWs) in Uttar Pradesh has increased dramatically from about three thousand in 1951 to 600 thousand in 1977 and 1.05 million by March 1980. In fact, by the mid-1970s, tube well irrigation had surpassed canal irrigation, which had previously been the dominant mode of irrigation. In the state of Uttar Pradesh, there were 21.1 PTW per 100 hectares on average. With the exception of North West India, this compares favourably to other South Asian regions.

The author noticed that the lower castes were gaining ground on the upper castes in terms of owning such implements. However, owning mechanical water extraction devices and modern agricultural implements is still out of reach for marginal farmers, particularly SCs/STs. Despite the apparent success of the free boring schemes, this is the case. In their paper 'Food Processing and Contract in AP: A Small Farmer Perspective,' Mahendra Dev and Rao (2005) 44 claim that the contract system is working well and has solved the problems of marketing, input purchase, and extension services. They also suggested that the contracts be improved by completing the grading process solely at the collection centre. They noticed that the grading processes are repeated at the factory, and that the shrivelled fruits are rejected this time.

According to them, this results in a weight variation of up to 10%-20%. They want the grading to be done on-site at the collection centre. Some respondents (15%) believed that the drip subsidy could be extended in order to boost crop production.

Around 75% of farmers wanted the government to provide power for at least 10 hours per day, rather than the current 6-7 hours. Crop insurance and high-quality pesticides were also requested. In his paper 'Food Security, Agrarian Crisis, and Rural Livelihood – Implications of Women,' Krishnaraj (2006) emphasises the role of agricultural growth in alleviating poverty and increasing farmer per capita income and food grain availability across India. According to him, the all-India consumer price index for agricultural and rural labourers, particularly for food, shows a significant increase in the cost of living.

This is the result of government policies that favour rice and wheat growth by promoting hybrid varieties, high input technology, and a focus on irrigated areas.Aside from a decrease in the area under cultivation of nutritionally rich coarse cereals, their production and yield have decreased due to a lack of support. Recent data on lower food consumption and diversion to non-food items is aggregating poor farmers' financial problems, resulting in lower incomes due to poor agricultural returns. The average wage growth rate has slowed from 0.696 in the late 1970s to 0.29 in 2003-04.

Using the OLS Method, M. Ghose (2007) investigated the effect of agricultural development, agrarian structure, and some other variables on rural poverty in his paper 'Agricultural Development, Agrarian Structure, and Rural Poverty.'

In terms of agricultural production per head of rural population, he discovered that the incidence of rural poverty is inversely proportional to agricultural development, implying the existence of a trickle-down effect in rural India. The findings suggest that rural poverty can be significantly reduced by increasing productive employment in rural areas and maintaining a reasonable rural wage rate.

As a result, any increase in employment in the agricultural and non-agricultural sectors will help to alleviate rural poverty. In their article 'Progress and Problems in Agricultural Insurance,' Raju and Chand (2007) examined the various government-sponsored agricultural insurance schemes and their impact on farmers' well-being. Agricultural insurance, they claim, has served little purpose despite various schemes introduced in the country from time to time. The coverage in terms of area, number of farmers, and value of agricultural output is very limited; indemnity payments based on the "area approach" miss affected farmers outside the compensated area, and the majority of the schemes are unviable.This will necessitate renewed government efforts in terms of developing appropriate mechanisms and providing financial assistance for agricultural insurance. Providing similar assistance to private sector insurance would aid in increasing insurance coverage and improving the long-term viability of insurance schemes.

In the study Bhatnagar, V., &Poonia, R. C. (2022) states that agriculture is the most important source of income for humans, animals, and all other living things. Agriculture is also very important in India's economy, and it describes the importance of agriculture as well as the factors that influence its development. This chapter discusses the international agricultural scenario, the current state of Indian agriculture, and the agricultural position of Rajasthan (Indian state). This chapter discusses total production, total imports and exports, irrigation method, net irrigation area, types of crops, fertilisation consumption, and the highlights of the Union Budget 2018-19 for Indian agriculture. This chapter also covers the geography of Rajasthan in terms of agriculture, crop production, and fertiliser consumption. The study came to a conclusion with some recommendations for the future.

The study of Chand, R., Joshi, P., &Khadka, S. (2022) brings together a variety of viewpoints on the transformational change that India's agriculture and related sectors require. The study promotes approaching this change through eight broad areas, indicating the policy shifts needed to meet the challenges of the coming decade.

Highlighting the importance of thinking for a post-Green Revolution future (2021-2030). A total of ten contributions are included in the study. There are eight thematic issues, including transforming Indian agriculture, dietary diversity for nutritive and safe food; climate crisis and risk management; water in agriculture; pests, pandemics, preparedness, and biosecurity natural farming; agro ecology and biodiverse futures; science, technology, and innovation in agriculture; and structural reforms and governance.

Objectives of the Study

The present research paper is designed to focus on the objectives; which can be stated as follows

- To list out the various sustainable systems and practices in Indian Agriculture

- To evaluate the impact of sustainable agriculture practices

Various Sustainable System and Practices in Indian Agriculture

As per the report of Sustainable Agriculture in India 2021, it is discovered that there are 30 sustainable agriculture practises (SAPs) are used in Indian farming sector. Among these 30 practices, some of the practices are only focuses on one aspect of agriculture. Others, on the other hand, are more holistic when it comes to agriculture in general. In this report, these are all referred as "sustainable agriculture practises and systems" as a group (SAPSs). Many of these practises overlap, and Table 1 and Table 2 list out all these practices. together a variety of viewpoints on the transformational change that India's agriculture and related sectors require. The study promotes approaching this change through eight broad areas, indicating the policy shifts needed to meet the challenges of the coming decade, highlighting the importance of thinking for a post-Green Revolution future (2021-2030).

Table 1: Sustainable Agriculture Systems in India

System	Area under the system/ practice (million ha)	Scale of adoption (number of farmers in millions)	Geographical spread (number of states)
Premaculture	< 0.05	0.01	3 to 4
Organic Farming	2.8	1.9	All
Natural Farming	0.7	0.8	3 to 4
System of rice intensification (SRI)	3	> 3	25
Biodynamic agriculture	0.1	0.1	10
Conservation agriculture	2^2	1	4 to 5
Integrated farming system (IFS)	< 0.1	< 0.1	10 to 15
Agroforestry	25	< 5	All
Integrated pest management (IPM)	5	5	22
Precision farming	9.2	3	24
Silvipastoral systems	NA	NA	NA
Vertical farming	NA	NA	NA
Hydroponics/Aeroponics	NA	NA	NA
Crop-livestock-fisheries farming system	NA	NA	NA

Source: Sustainable Agriculture in India 2021

Table 2: Sustainable Agriculture Practices in India

System	Area under the system/practice (million ha)	Scale of adoption (number of farmers in millions)	Geographical spread (number of states)
Vermicomposting	3.5	1.5	ALL
Drip irrigation/Sprikler	NA	NA	NA
Crop rotation	30^3	15	ALL
Intercropping	1^6	0.8	ALL
Cover crops	1.9^5	1.5	ALL
Mulching	20	< 5	17
Contour farming	2	3	19
Rainwater harvesting- artificial recharge of groundwater	$>20^4$	< 5	All
Floating farming	0	0	1
Plastic mulching	NA	NA	NA
Shade net house	NA	NA	NA
Alternate wetting and drying technique (for rice)	NA	NA	NA
Saguna rice technique	NA	NA	NA
Farm pond lined with plastic film	NA	NA	NA
Direct seeding of rice	NA	NA	NA
Canopy management	NA	NA	NA
Mangrove and non-Mangrove bio-shields	NA	NA	NA

Source: Sustainable Agriculture in India 2021

Crop rotation, one of the most basic SAPSs, is the most widely used in the country, with 30 million hectares and 15 million farmers. Practices such as agroforestry and rainwater harvesting, which have received a lot of attention in national programmes, are also getting more attention. Despite the fact that agroforestry covers a large area, it is primarily used by large cultivators. There is very little documented information on the prevalence of mulching; however, one stakeholder estimated that it covers about 20 million hectares.

Although the area under precision farming (nine million ha) appears to be large, it primarily consists of the area under micro-irrigation, which is a component of precision farming. The National Mission on Micro Irrigation has made significant progress in promoting micro-irrigation in the country over the years. Despite decades of promotion, Integrated Pest Management has a low coverage of only 5 million hectares. Intercropping is more common in the south and west of the country, covering nearly one million hectares. However, due to a lack of reliable estimates, the estimate excludes intercropping areas in horticultural crops.

Organic farming currently covers only 2% of the country's total net sown area, despite government policy support (140 million ha). There are about two million certified organic farmers in India, but there is no reliable information on uncertified organic farmers. Biodynamic agriculture, a type of organic farming, is thought to cover 0.1 million hectares (where biodynamic inputs are explicitly used along with organic farming practices). In the last two to three years, natural farming has seen a faster rate of adoption. Natural farming is practised by nearly one million farmers, mostly in Andhra Pradesh, Karnataka, Maharashtra, and Himachal Pradesh. As it has been primarily popular among small and marginal farmers, the associated area is approximately 0.7 million ha. In the last five years, the popularity of the rice intensification system has exploded, with an estimated area of around 3 million ha across the country. Around 2 million hectares of land are under partial conservation agriculture, mostly in a few states in the Indo-Gangetic Plains.

Impact of Sustainable Agriculture Practices

In general, environmental balance, low cultivation costs, a clean environment, and nutritious food free of pesticide residues are the main benefits of sustainable agriculture. Sustainable agriculture often includes a wide range of production methods, including both conventional and organic methods. Long-term results are expected from a regionally integrated system of plant and animal production practises. However, the impact of sustainable agricultural practices on economic, ecological and social aspects are enlisted as follows

Economic Impact

In terms of geographical coverage and the number of long-term assessments, the evidence on SAPSs' impact on farmer incomes remains insufficient. Despite this critical limitation, the literature suggests that a few SAPSs have the potential to increase income by lowering production costs (CA, natural farming), diversifying agricultural production (IFS, intercropping), and charging premium prices (organic produce).

Despite the conceptual limitations of accurately estimating farm productivity, we discover some emerging yield patterns under a few SAPSs. Organic farming yields are lower than conventional farming, at least in the short term (2-3 years). After this time period, some studies show that some crops can produce equal or even higher yields, especially once the soil form and structure have evolved after a few years of biological inputs. For most crops, short-term studies of natural farming show no statistically significant changes in yields. The effects of SRI on yield have been well documented, with statistically significant increases in various paddy varieties. Vermicomposting, agroforestry, and crop diversification are examples of resource-saving practises that have improved yields. However, because there are few studies on the long-term effects of SAPSs on yields, it's difficult to extrapolate conclusions.

Ecological Impact

The impact of various SAPSs on water-use efficiency has been documented in a number of studies.

Water conservation has been aided by SRI, CA, precision farming, rainwater harvesting, contour farming, cover crops, mulching, crop rotation, and agroforestry. Smallholder farmers like rainwater harvesting and SRI because they are simple to implement.

In drought-prone Andhra Pradesh, pre-monsoon dry sowing in natural farming is regarded as a breakthrough that merits further investigation.

Agroforestry, SRI, and CA are the SAPSs with the most evidence for climate mitigation. The ability of agroforestry to sequester carbon (above and below ground) has been well documented. The SRI appears to promote aerobic soil conditions, which reduce methane emissions, according to a growing body of evidence. Intermittent irrigation, which is an integral part of SRI, can, however, increase nitrous oxide emissions. Overall, the SAPSs' long-term carbon sequestration impacts in India must be assessed.

Agroforestry, IFS, permaculture, natural farming, organic farming, conservation agriculture, and crop diversification strategies (rotation, intercropping, mixed) are some of the SAPSs that increase the spatial, vertical, and temporal diversity of species on a farm (and landscape) level. While research articles mention the impact on biodiversity, there are no studies that provide solid empirical evidence.

The anecdotal evidence that various SAPSs have positive health effects, primarily due to dietary diversity and reduced exposure to harmful chemicals like pesticides. There are no empirical studies comparing SAPSs to conventional agriculture in terms of health outcomes.

Social Impact

In Indian agriculture, women account for more than 70% of the workforce. However, there are few studies that focus on the gender outcomes of SAPSs. Women's roles are defined by a few practises such as vermicomposting, organic farming, IFS, and rainwater harvesting, but evidence on their impact is lacking.

To fully comprehend the impact of various SAPSs on women's workloads, income, empowerment, and employment, more research is required.

Conclusion

While states like Sikkim and Andhra Pradesh are leading the way in India in terms of sustainable agriculture, adoption is still on the fringes across the country. Similarly, there is a scarcity of evidence on the impact of its outcomes on the economic, social, and environmental fronts. On the one hand, gathering more long-term evidence is utmost important. In addition, existing evidence should be used to scale up context-specific SAPSs. Rainfed areas could be the first to scale up because they already practise low-resource agriculture, have low productivities, and stand to benefit the most from the transition. Farmers in irrigated areas will follow suit as positive results at scale emerge.Significantly increase budgetary allocation to sustainable agriculture, allowing for evidence-based scale-up across the country. Focus on regional and practice-specific priorities at the tactical level, which range from technological innovation to help mechanise labour-intensive processes to farmers' capacity building in knowledge-intensive practises. Finally, shift the focus of national policy away from food security and toward nutrition and total farm productivity. It would help to recognise the critical role that sustainable agriculture can play in ensuring India's food security in a world where climate change is a reality.

References

- AakashSunaratiya and Rupali P. Kapgate ,(2021), Organic Farming : Key of Sustainable Agriculture, Agriculture & Food : E-Newsletter ,Volume 3 - Issue 6 - June 2021.
- AnaghaBhilare, (2013), Sustainable Agriculture and Land Management, Research India Publications, International Journal of Agriculture and Food Science Technology. Volume 4, Number 9, pp. 881-886.

- Bhatnagar, V., &Poonia, R. C. (2022). Sustainable development in agriculture: past and present scenario of Indian agriculture. In Research Anthology on Strategies for Achieving Agricultural Sustainability (pp. 1342-1364).IGI Global.
- Braat, L. (1991), The predictive meaning of sustainability indicators. In search of Indicators of Sustainable Development (O. Kurk and H. Vergruggen, Eds.). Dordrecht Kluwer Academic Publishers, The Netherlands.
- Chand, R., Joshi, P., &Khadka, S. (2022). Indian Agriculture Towards 2030: Pathways for Enhancing Farmers' Income, Nutritional Security and Sustainable Food and Farm Systems.
- D.M. Hegde and S.N. SudhakaraBabu, Sustainable Agriculture, krishi.icar.gov.i n/jspui.
- Dr V. Basil Hans, (2013), An Analysis of Sustainable Agricultural Development in India, Technical Report for the Post – 2015, Development Agenda, 18 September 2013.
- Dr. Babar Someshwar Narayan,(2012), Sustainable Agricultural Development and Organic Farming in India, Golden Research Thoughts ,Vol.1,Issue.XI/May 2012 pp.1-4.
- Gupta, Niti, Shanal Pradhan, Abhishek Jain, and Nahya Patel. 2021. Sustainable Agriculture in India 2021: What We Know and How to Scale Up. New Delhi: Council on Energy, Environment and Water.
- Mac Rae, et.al,(1990), Policies, programs and regulations to support the transition to sustainable agriculture in Canada, American Journal of Alternative Agriculture, 5(2): 76.
- Saroj Kumar Singh and AnkitaParihar,(2015), Challenges of Sustainable Agriculture Development in India, Journal of Agroecology and Natural Resource Management, KrishiSanskriti Publications, Volume 2, Issue 5;pp. 355-359.
- Simon Bell and Stephen Morse,(2008), Eds. Sustainability Indicators: Measuring the Immeasurable? Second edition.Earthscan Publishing. London
- Wezel, A., &Soldat, V. (2009)."A quantitative and qualitative historical analysis of the scientific discipline of agroecology", International Journal of Agricultural Sustainability, 7(1), 3– 18.https://doi.org/10.3763/ijas.2009.0400.

- Golden Research Thoughts ,Vol.1,Issue.XI/May 2012 pp.1-4.
- Gupta, Niti, Shanal Pradhan, Abhishek Jain, and Nahya Patel. 2021. Sustainable Agriculture in India 2021: What We Know and How to Scale Up. New Delhi: Council on Energy, Environment and Water.
- Mac Rae, et.al,(1990), Policies, programs and regulations to support the transition to sustainable agriculture in Canada, American Journal of Alternative Agriculture, 5(2): 76.
- Saroj Kumar Singh and AnkitaParihar,(2015), Challenges of Sustainable Agriculture Development in India, Journal of Agroecology and Natural Resource Management, KrishiSanskriti Publications, Volume 2, Issue 5;pp. 355-359.
- Simon Bell and Stephen Morse,(2008), Eds. Sustainability Indicators: Measuring the Immeasurable? Second edition.Earthscan Publishing. London
- Wezel, A., &Soldat, V. (2009)."A quantitative and qualitative historical analysis of the scientific discipline of agroecology", International Journal of Agricultural Sustainability, 7(1), 3–18.https://doi.org/10.3763/ijas.2009.0400.

Chapter 7   **ISBN: 978-81-948672-7-2**

An Analysis of Sustainable Development of Agriculture Sector in Karnataka: An Overview

DR. SIDDAPPAJI .D
Faculty
Department of Economics
Mandya University , Mandya

DR. ANAND .C
Faculty
Department of Economics
Mandya University , Mandya

Email: *anand1985mysore@gmail.com*

Abstract: The present research article explored that the sustainable agriculture is farming in sustainable ways meeting culture's present food and textile needs, without compromising the ability for future generations to meet their needs. It can be based on an accepting of eco-system services in the agriculture development. There are many methods to increase the sustainability of agriculture sector. Sustainable agriculture is one of the ancient, primary occupations of Karnataka.It is the main source of livelihood for many people. It is the backbone of our state economy. Economic progress depends on agriculture.It provides to food for the people and raw materials to industries.

The research paper discussed the sustainable development of agriculture sector in Karnataka state of India. This study analyzes the area, production, and productivity of agriculture sector in Karnataka and to identify the social factors of sustainable agriculture sector. The growth in the area and production of different major crops in Karnataka was estimated using the compound growth function and correlation. The necessary secondary data were collected for a period of 6 years during period from 2015-16 to 2020-21. The research study was statistical tools used like Percentages and Compound Annual Growth Rate (CAGR) and Correlation were analyzed through excel and SPSS 21 statistical software. Growth rates showed a significant positive growth in area total cereals, pulses, food grains, oilseeds, and commercial crops showed significant positive growth.

Keywords: Agriculture, Components of Sustainable Agriculture, Major Crops, Area, and Production.

Introduction:

A sustainable agriculture approach seeks to utilize natural resources in such a way that they can regenerate their productive capacity, and also minimize harmful impacts on ecosystems beyond a field's edge. It can be based on an understanding of ecosystem services. Agriculture sector has a vast environmental footprint, playing a major role in causing climate change, water scarcity, water pollution, land degradation, deforestation and other processes; it is simultaneously, causing environmental changes and being impacted by these changes. Sustainable agriculture consists of environment friendly methods of farming that allow the production of crops or livestock without damage to human or natural systems. It involves preventing adverse effects to soil, water, biodiversity, surrounding or downstream resources as well as to those working or living on the farm or in neighboring areas.

Elements of sustainable agriculture can include permaculture, agroforestry, mixed farming, multiple cropping, and crop rotation. The elasticity of India's agriculture sector can be seen from the fact that despite the COVID-19 pandemic, its performance in output was strong. As per Census 2011, about 30 per cent of the total workers in the state are still engaged in agricultural and allied sector activities, which accounts for about 8.73 per cent of the Karnataka's Gross State Value Added (GSVA) for the year 2019-20 (at constant prices).The main crops grown are Rice, Ragi, Jowar (sorghum), maize, and pulses (Tur and gram) in addition to oilseeds and a number of other cash crops. Cashews, coconut, are canut, cardamom, chillies, cotton, sugarcane and tobacco are also produced. Karnataka is the largest producer of coarse cereals, coffee, raw silk and tomatoes among the states in India. Horticultural crops are grown in an area of 16,300 km2 and the annual production is about 9.58 million tons. The income generated from horticulture constitutes over 40% of income generated from agriculture and it is about 17 per cent of the state's GDP.

Objectives of the study:

1. To study the performance and component of sustainable agriculture development in Karnataka.
2. To analyze the current growth in trends of area and production of agriculture sector in the state.

Hypothesis of the study:

1.There is a significant relationship between area and production of the sustainable agriculture sector in Karnataka.

Research methodology:

The present paper is mainly based on secondary data information. The secondary data has been collected from various annual reports of Central Statistics Office, Ministry of Statistics and Programme Implementation, Government of India, Minister of Agriculture and Farmers Welfare, Economic Survey of India and Karnataka (2020-21), Directorate of Economics and Statistics, journals, articles etc. The research paper was study period of 2015-16 to 2019-20. The study was statistical tools used like Percentages and Compound Annual Growth Rate (CAGR) and Correlation were analyzed through excel and SPSS 21 statistical software. In Karnataka, agriculture has occupied about 12.31 million hectares of land, which includes 64.6 per cent of the total area. The main season for sustainable agriculture in the state is monsoon as irrigation is done in 26.5 per cent of the total cropped area. The principal cropping pattern of the region is influenced not only by agro-climatic conditions like rainfall, soil, and temperature, etc. but also by State Government policies and schemes for crop production in the form of subsides, tariff and speed of infrastructure development.

Components of sustainable agriculture:

The major components of sustainable agriculture farming are - soil management, crop management, water management, disease/pest management, and waste management. It is the methods used that are often fundamentally different.

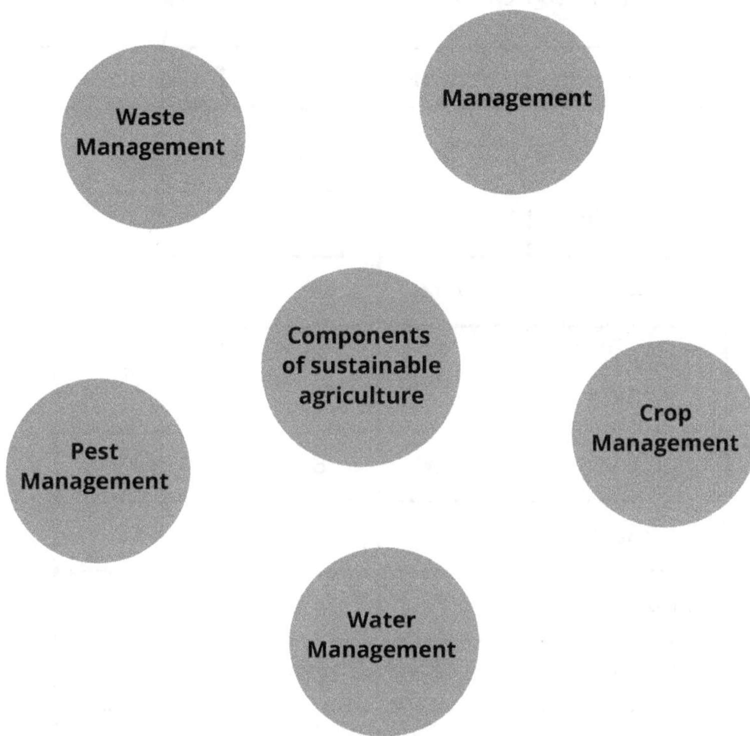

Chart:1 Key Components of Sustainable Agriculture Sector

Growth in Trends of Major Crops: Area and Production

The 2nd Advance Estimates of Production (2020-21) based on area coverage under major crops during kharif, rabi and summer seasons, loss of crops due to excess rains/floods in some parts indicate likely production of 117.38 lakh tonnes cereals and 19.28 lakh tonnes of pulses against the target of 110.02 and 23.03 lakh tonnes, respectively. Oilseeds production is estimated to be 11.46 lakh tonnes against the target of 13.05 lakh tonnes. Production of cotton is likely to be 19.11 lakh bales against the target of 13.90 lakh bales.

TABLE- 1: Season- wise area of principal crops in karnataka from 2017-18 to 2020-21(Unit: Area in Lakh Hectares)

Crop/Group	2017-18	2018-19	2019-20	2020-21
kharif				
Cereals	30.00	30.46	35.04	35.65
Pulses	16.02	16.34	22.49	22.06
Total Food grains	46.02	46.80	57.53	57.71
Oilseeds	8.27	8.91	8.93	9.46
Cotton	5.21	4.71	8.03	7.50
Sugarcane	4.00	4.00	4.29	4.50
Tobacco	0.95	0.93	0.95	0.70
Annual	**110.47**	**112.15**	**137.26**	**137.58**
Rabi				
Cereals	13.60	10.45	10.55	8.84
Pulses	14.16	12.40	10.23	9.92
Total Food grains	27.76	22.85	20.78	18.76
Oilseeds	0.97	1.48	0.55	0.78
Cotton	0.26	0.07	0.14	0.15
Sugarcane	-	-	-	-
Tobacco	-	-	-	-
Annual	56.75	47.25	42.25	38.45

Summer			
Cereals	2.64	1.72	3.26	2.60
Pulses	0.057	0.18	0.05	0.08
Total Food grains	2.697	1.90	3.31	2.68
Oilseeds	1.75	2.05	1.18	1.85
Cotton	-	-	-	-
Sugarcane	-	-	-	-
Tobacco	-	-	-	-
Annual	**7.144**	**5.85**	**7.8**	**7.21**

Source: Government of Karnataka(2020), Department of Agriculture, Planning Programme Monitoring and Statistics Department, Bengaluru – 2020-21.

The above Table –1 gives the information about the season-wise area of principal crops in Karnataka from 2017-18 to 2020-21. Under the Kharif season, the area of major crops in the agriculture sector is 11047 lakh ha in 2017-18, which has increased to 137.58 lakh ha in 2020-21.   During the Rabi season, during 2017-18, the area of major crops in the agriculture sector is 56.75 lakh ha, which has decreased to 38.45 lakh ha in 2020-21. During the summer season, the area of major crops in the agriculture sector increased trends from 7.144 lakh ha in 2017-18 to 7.21 lakh ha in 2020-21.

TABLE - 2: Season- wise production of principal crops in Karnataka from 2017-18 to 2020-21 (Unit: Lakh Tonnes, Cotton in Lakh bales of 170 Kgs in lint form)

Crop/Group	2017-18	2018-19	2019-20	2020-21
Kharif				
Cereals	93.00	70.77	90.38	99.38
Pulses	11.42	8.32	14.07	13.18
Total Food grains	104.42	79.09	104.45	112.56
Oilseeds	10.02	7.39	8.59	9.18
Cotton	18.25	9.21	23.19	19.11
Sugarcane	315.00	322.15	360.34	410.40
Tobacco	0.89	0.73	0.83	0.54
Annual	**553.00**	**497.66**	**601.85**	**664.27**
Rabi				
Cereals	18.48	10.78	14.67	9.93
Pulses	10.67	5.88	7.47	6.08
Total Food grains	29.15	16.66	22.14	16.01
Oilseeds	0.73	1.25	0.49	0.52
Cotton	0.19	0.06	0.10	0.08
Sugarcane	-	-	-	-
Tobacco	-	-	-	-
Annual	**59.22**	**34.63**	**44.87**	**32.62**

Summer				
Cereals	8.12	5.06	9.81	8.07
Pulses	0.017	0.006	0.011	0.02
Total Food grains	8.137	5.12	9.82	8.09
Oilseeds	2.04	2.03	1.32	1.76
Cotton	0.06	-	-	-
Sugarcane	-	-	-	-
Tobacco	-	-	-	-
Annual	**18.374**	**12.216**	**20.961**	**17.94**

Source: Government of Karnataka(2020), Department of Agriculture, Planning Programme Monitoring and Statistics Department, Bengaluru – 2020-21.

The above Table –2 indicates the season-wise production of principal crops in Karnataka from 2017-18 to 2020-21. During the Kharif season, the annual crop predominately increasing production of principal crops in the agriculture sector is 553.00 in 2017-18, which has grown to 664.27 in 2020-21. In the Rabi season, the annual production of major crops is 59.22 in 2017-18, which sharply decreased to 32.62 in 2020-21. During the summer season of 2017-18, the annual major crops of the agriculture sector are 18.374, this decreased to 17.94 in 2020-21.

Table- 3: Area of Principal Crops in Karnataka from 2017-18 to 2020-21(Unit: Area in Lakh Hectares)

Crop	2015-16	2016-17	2017-18	2018-19	2019-20	CAGR (%)
Rice	11.10	10.34	9.93	11.99	12.48	3.90
Jowar	11.04	9.48	10.88	9.94	9.14	-3.25
Ragi	7.05	5.98	7.78	5.55	6.74	-1.63
Maize	12.20	13.70	13.07	14.09	15.00	4.51
Bajra	1.66	2.42	2.32	1.94	3.39	12.83
Wheat	1.74	1.68	1.93	1.58	1.58	-2.15
Minor millets	0.28	0.21	0.37	0.19	0.52	12.05
Total cereals	**45.07**	**43.18**	**46.27**	**45.28**	**48.85**	**1.96**
Tur	6.57	12.14	8.85	15.16	16.26	22.92
Total pulses	28.31	29.66	30.26	35.70	32.77	4.90
Total Food grains	73.38	73.47	76.53	80.98	81.62	3.15
Ground nut	5.70	6.66	5.65	5.41	5.30	-3.47
Total Oil seeds	12.86	12.93	11.00	9.99	10.66	-6.14
Cotton	6.42	5.10	5.47	7.18	8.17	8.59

Sugarcane	6.02	4.88	5.33	7.94	7.65	10.14
Tobacco	0.84	0.90	0.95	0.88	0.95	2.26
Areca nut (Processed)	2.48	2.55	2.79	4.76	5.00	22.46
Coconut	4.41	4.37	4.46	6.10	6.10	10.32
Dry chillies	1.02	1.28	1.02	1.58	0.74	-4.22
Pepper	0.35	0.38	0.41	1.48	1.61	55.45
Cardamon	0.18	0.17	0.15	0.02	0.02	-47.98

Source: Government of Karnataka(2020), Department of Agriculture, Planning Programme Monitoring and Statistics Department, Bengaluru – 2020-21.

Table –2 indicates the season-wise production of principal crops in Karnataka from 2017-18 to 2020-21. During the Kharif season, the annual crop predominately increasing production of principal crops in the agriculture sector is 553.00 in 2017-18, which has grown to 664.27 in 2020-21. In the Rabi season, the annual production of major crops is 59.22 in 2017-18, which sharply decreased to 32.62 in 2020-21. During the summer season of 2017-18, the annual major crops of the agriculture sector are 18.374, this decreased to 17.94 in 2020-21.

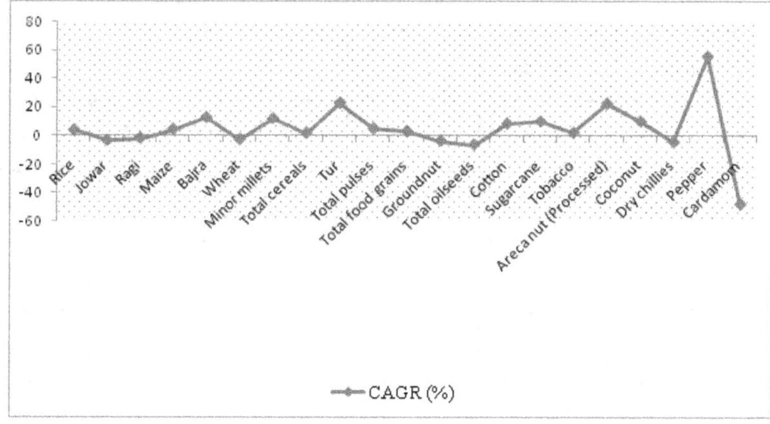

The area of principal crops in the agriculture sector during the period from 2015-16 to 2019-20, in terms of lakh ha, is presented in Table -3 and Graph -3. It is clear that the major crops like total cereals, pulses, food grains, oilseeds, cotton, sugarcane, and tobacco. During 2015-16, the total cereals area was 45.07 lakh ha, which predominantly increased to 48.85 lakh ha in 2019-20. The total pulses area is 28.31 lakh ha in 2015-16, which has increased to 32.77 lakh ha in 2019-20. In 2015-16, the total food grains area was 73.38 lakh ha, which increased to 81.62 lakh ha in 2019-20. The total oil seed of area position is 12.86 lakh ha in 2015-16, which increased to 12.93 lakh ha in 2016-17, and then it has decreased to 10.66 lakh ha in 2019-20. The above Table also shows the CAGR for area of major crops in the agriculture sector in the state during the period between 2015-16 and 2019-20. The CAGR for the major crops of total cereals is 1.96 per cent, followed by the total pulses (4.90%), total food grains (3.15%), total oil seeds (-6.14%), cotton (8.59%), sugarcane (10.14%), tobacco (2.26%) and other commercial crops showed in above Table-3 and Graph -1.

Table –4

Production of Principal Crops in Karnataka from 2017-18 to 2019-20

(Unit: Lakh Tonnes, Cotton in Lakh bales of 170 Kg in Lint form)

Crop	2015-16	2016-17	2017-18	2018-19	2019-20	CAGR (%)
Rice	30.21	28.74	29.40	34.58	39.48	7.47
Jowar	7.96	8.46	12.01	9.31	10.45	6.61
Ragi	11.88	4.91	15.41	6.78	11.62	2.82
Maize	33.10	33.14	55.52	37.77	47.46	8.89
Bajra	1.11	2.55	3.53	1.77	3.67	22.46
Wheat	1.56	1.71	2.24	1.63	1.80	2.41

Minor millets	0.10	0.07	0.25	0.16	0.38	41.86
Total cereals	**85.92**	**79.85**	**118.38**	**92.00**	**114.86**	**7.49**
Tur	2.42	12.14	8.13	9.47	11.26	32.69
Total pulses	10.53	20.41	21.53	18.46	21.55	14.27
Total Food grains	96.44	99.99	139.91	110.46	136.41	8.25
Groundnut	3.95	4.19	5.99	3.97	5.03	4.39
Total Oil seeds	7.09	8.05	12.20	7.92	10.41	-6.14
Cotton	11.52	10.24	17.71	14.00	23.29	18.78

Crop	2015-16	2016-17	2017-18	2018-19	2019-20	CAGR (%)
Sugarcane	363.14	273.38	374.61	424.11	360.34	4.33
Tobacco	0.49	0.65	0.95	0.71	0.83	12.10
Areca nut Processed	4.86	5.17	5.10	8.51	10.28	23.36
Coconut	3688.82	3417.20	3284.69	4321.76	5030.77	8.93
Dry chillies	1.03	2.60	2.07	1.96	1.29	1.69
pepper	2.68	2.29	2.57	9.41	6.30	36.65
Cardamom	0.10	0.02	0.01	0.002	0.003	-37.57

Source: Government of Karnataka (2020), Final Estimates of Directorate of Economics & Statistics – 2015 to 2019.

Note: *Sugarcane production for harvested area of 4.00 lakh ha.in 2017-18, 4.96 lakh ha, in 2018-19 & 4.29 lakh ha, in 2019-20.

Table –2: CAGR of Production of Principal Crops in Karnataka from 2017-18 to 2019-20

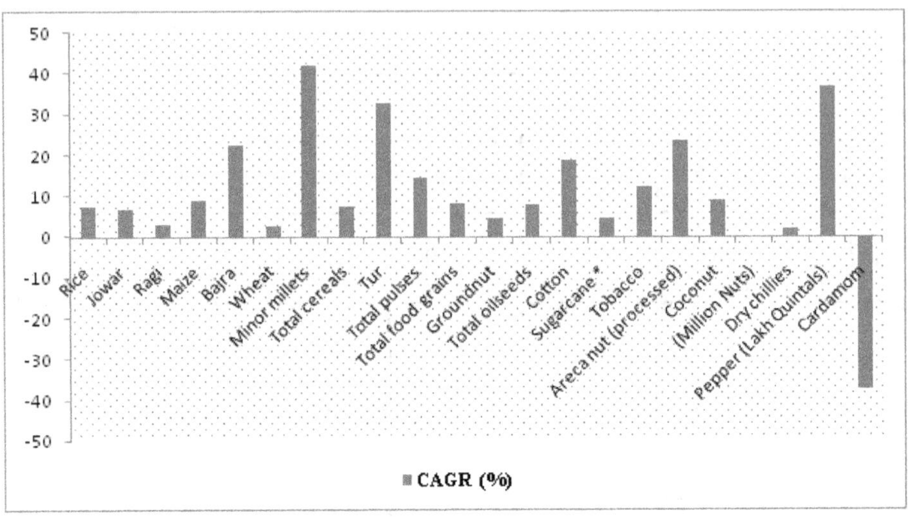

Table – 4 and Graph – 2 shows the production of principal crops in the agriculture sector during the period from 2015-16 to 2019-20, in terms of lakh tonnes. It is clear from the above table, the production of principal crops in the agriculture sector like total cereals, pulses, food grains, oilseeds, cotton, sugarcane, and tobacco. During 2015-16, the production of total cereals was 85.92 lakh tonnes, which significantly increased to 114.86 lakh tonnes in 2019-20. The total pulses production was 10.52 lakh tonnes in 2015-16, which has increased to 21.53lakh tonnes in 2017-18; then it has decreased to 18.46 lakh tonnes in 2018-19; and then it has sharply increased to 21.55 lakh tonnes in 2019-20. In 2015-16, the production of total food grains was 96.44 lakh tonnes, which increased to 136.41 lakh tonnes in 2019-20. The total oilseed of production position was 7.09 lakh tonnes in 2015-16, which increased to 12.20 lakh tonnes in 2017-18; then it has decreased to 7.92 lakh tonnes in 2018-19; and then it has sharply increased to 10.41 lakh tonnes in 2019-20. The commercial major crops like cotton, sugarcane, tobacco, etc. The total production of cotton was 11.52 lakh bales in 2015-16, which significantly increased to 23.29 lakh bales in 2019-20.

The tobacco production was 0.49 lakh tonnes in 2015-16; which increased to 0.83 lakh tonnes in 2019-20. The above table also illustrations the CAGR for production of principal crops in the agriculture sector during the period between 2015-16 and 2019-20. The CAGR for the production of principal crops of total cereals is 7.49 per cent, followed by the total pulses (14.27%), total food grains (8.25%), total oilseeds (7.81%), cotton (18.78%), sugarcane (4.33%), tobacco (12.40%) and other commercial crops showed in above Table – 4 and Graph - 4.

Testing Hypothesis

H0: There is no significant relationship between area and production of the sustainable agriculture sector in Karnataka.

H1: There is a significant relationship between area and production of the sustainable agriculture sector in Karnataka.

Table- 5
Correlation Between Area and Production of Agriculture Sector in Karnataka

Variables	Mean	Std. Deviation	Pearson Correlation	Sig.
Agriculture Area	204.8400	10.45699	0.647*	0.238
Agriculture Production	617.6660	82.21006		

*. Correlation is significant at the 0.05 level (2-tailed).

Table – details of the correlation between area and production of agriculture sector in Karnataka. The calculated mean values of agriculture area and agriculture production are 204.8400 and 10.45699 respectively. The tested standard deviation values of agriculture area and agriculture production are 617.6660 and 82.21006 respectively.

The tested Pearson correlation value is .647, at significant 10% level. However, the null hypothesis accepted and the alternative rejected. Hence, it implies that there is a significant relationship between agriculture area and production of agriculture sector in Karnataka.

Conclusion

Karnataka state is considered a tiny part of India, as it reveals significant features of the country in climate, soil types, rainfall, crops grown, and a variety of natural resources. It has ideal soil and climate suited for raising a different variety of principal crops; tropical region, temperate region, humid and arid regions are grown in the state. Karnataka economy is one of the furthermost important attributes of the agriculture sector. Sustainable agriculture is considered to be one of the primary occupations for the inhabitants in the state. In Karnataka, the majority of the people are involved in growing crops especially in the rural areas. Sustainable agriculture is considered to be one of the major occupations for the inhabitants of the state. The majority of the people in the state are engaged in growing crops, particularly in rural areas. The present research papers analyze the growth in trends of area and production of the agriculture sector and it has find out the relationship between area and production of agriculture in the state. This study found from the correlation between area and production of agriculture sector in Karnataka. The calculated mean values of agriculture area and agriculture production are 204.8400 and 10.45699 respectively. The tested standard deviation values of agriculture area and agriculture production are 617.6660 and 82.21006 respectively. The tested Pearson correlation value is .647, at significant 10% level.

References:

- Bowler, I. (2002). Developing sustainable agriculture. Geography, 205-212.
- Charles, N. A. (2019). Assessment of Watersheds for Sustainable Agriculture in Karnataka, India.
- Debnath, S., Adamala, S., Palakuru, M., &Debnath, S. (2020). An overview of Indian traditional irrigation systems for sustainable agricultural practices. Int J Mod Agric, 9, 12-22.
- Government of Karnataka (2020), Department of Agriculture, Planning Programme Monitoring and Statistics Department, Bengaluru – 2020-21.
- Government of Karnataka (2020), Final Estimates of Directorate of Economics & Statistics – 2015 to 2019.

- Harwood, R. R. (2020). A history of sustainable agriculture. In Sustainable agricultural systems (pp. 3-19). CRC Press.
- Ikerd, J. E. (1993). The need for a system approach to sustainable agriculture. Agriculture, Ecosystems & Environment, 46(1-4), 147-160.
- Jaishankar, N., Janagoudar, B. S., Kalmath, B., Naik, V. P., &Siddayya, S. (2014, May). Integrated farming for sustainable agriculture and livelihood security to rural poor. In International Conference on Chemical, Biological and Environmental Science. May (pp. 12-13).
- Lichtfouse, E., Navarrete, M., Debaeke, P., Souchère, V., Alberola, C., &Ménassieu, J. (2009). Agronomy for sustainable agriculture: a review. Sustainable agriculture, 1-7.

Chapter 8

ISBN: 978-81-948672-7-2

An Analysis of Environmental Impact on Agricultural Production in India-A Special Reference to Hassan District

GIRISH M.C.
Assistant Professor
Department of postgraduate studies in
Economics Govt. Arts Commerce &
Postgraduate college – Autonomous Hassan
Email: girishmc2009@gmail.com

Abstract: The environmental impact of agriculture varies based on the wide variety of agricultural practices employed around the world. Ultimately, the environmental impact depends on the production practices of the system used by farmers. The connection between emissions into the environment and the farming system is indirect, as it also depends on other climate variables such as rainfall and temperature. Weather is the condition of the atmosphere at a particular place and time. It is characterized by parameters such as temperature, humidity, rain and wind. Climate is the long term pattern of weather conditions for a given area. Climate change refers to a statistically significant variation in either the mean state of the climate or its variability, persisting for an extended period. The environmental impact of agriculture involves a variety of factors from the soil, to water, the air, animal and soil diversity, people, plants, and the food itself. Some of the environmental issues that are related to agriculture are climate change, deforestation, genetic engineering, irrigation problems, pollutants, soil degradation, and waste. Therefore this paper analyses the environmental factor lies its impacts on agriculture production system and its consequences on agrarian community as well as impact on food security in the country.

Keywords: Climate change, Environment, Agriculture, Production, Sustainable

Introduction

The environmental impact of agriculture varies based on the wide variety of agricultural practices employed around the world. Ultimately, the environmental impact depends on the production practices of the system used by farmers. The connection between emissions into the environment and the farming system is indirect, as it also depends on other climate variables such as rainfall and temperature. Weather is the condition of the atmosphere at a particular place and time. It is characterized by parameters such as temperature, humidity, rain and wind. Climate is the long term pattern of weather conditions for a given area. Climate change refers to a statistically significant variation in either the mean state of the climate or its variability, persisting for an extended period. India is home to extraordinary variety of climatic regions, ranging from tropical in the south to temperate and alpine in the Himalayan north, where elevated regions receive sustained winter snowfall.

The nation's climate is strongly influenced by the Himalayas and the Tar Desert. Four major climatic groupings predominate into which fall seven climatic zones which are defined on the basis of temperature and precipitation.Climate change is the most important global environmental challenge facing humanity with implications for natural ecosystems, agriculture & health. The perusal of general circulation models (GCM s) on climate change indicate that rising levels of greenhouse gases (GHGs) are likely to increase the global average surface temperature by 1.5-4.5°C over the next 100 years. The difference of average temperature between the last ice age and present climate is 6°C. This will raise sea-levels, shift climate zones pole ward, decrease soil moisture and storms. Global warming is predicted to affect agricultural production. Therefore this paper analyse the environmental factor lies its impacts on agriculture production system and its consequences on agrarian community as well as impact on food security in the country.

There are two types of indicators of environmental impact: "means-based", which is based on the farmer's production methods, and "effect-based", which is the impact that farming methods have on the farming system or on emissions to the environment. An example of a means-based indicator would be the quality of ground water that is affected by the amount of nitrogen applied to the soil. An indicator reflecting the loss of nitrate to groundwater would be effect-based.

The environmental impact of agriculture involves a variety of factors from the soil, to water, the air, animal and soil diversity, people, plants, and the food itself. Some of the environmental issues that are related to agriculture are climate change, deforestation, genetic engineering, irrigation problems, pollutants, soil degradation, and waste. Climate change and agriculture are interrelated processes, both of which take place on a worldwide scale. Global warming is projected to have significant impacts on conditions affecting agriculture, including temperature, precipitation and glacial run-off.

These conditions determine the carrying capacity of the biosphere to produce enough food for the human population and domesticated animals. Rising carbon dioxide levels would also have effects, both detrimental and beneficial, on crop yields. Assessment of the effects of global climate changes on agriculture might help to properly anticipate and adapt farming to maximize agricultural production. Although the net impact of climate change on agricultural production is uncertain it is likely that it will shift the suitable growing zones for individual crops. Adjustment to this geographical shift will involve considerable economic costs and social impacts.

At the same time, agriculture has been shown to produce significant effects on climate change, primarily through the production and release of greenhouse gases such as carbon dioxide, methane, and nitrous oxide. In addition, agriculture that practices tillage, fertilization, and pesticide application also releases ammonia, nitrate, phosphorus, and many other pesticides that affect air, water, and soil quality, as well as biodiversity.

Agriculture also alters the Earth's land cover, which can change its ability to absorb or reflect heat and light, thus contributing to radiative forcing. Land use change such as deforestation and desertification, together with use of fossil fuels, are the major anthropogenic sources of carbon dioxide; agriculture itself is the major contributor to increasing methane and nitrous oxide concentrations in earth's atmosphere.

Methodology

The methodology of the present study exclusively relieson the secondary data and information of national and state level."An Analysis of Environmental Impact on Agricultural Development in India" is estimated on the basis of available secondary sources such are Government publications, reports of ministry of agriculture, agricultural departments, various state level and national level journals and books.

Objectives of the study

- To understand about the rural agriculture production trend and agricultural environmental system.
- To know the agricultural share in to the economic development.
- To understand the environmental impact on agriculture.

Climate change and India

The Inter-Governmental Panel on Climate Change (IPCC) was established by United Nations Environment Programme (UNEP) and the World Meteorological Organization (WMO) to provide the world with a clear view on the current state of Climate Change and its potential environmental and socio-economic consequences. IPCC defines climate change as 'a change in the state of the climate that can be identified by changes in the mean and/or variability of its properties, and that persists for an extended period, typically decades or longer. It refers to any change in climate over time, whether due to natural variability or as a result of human activity'. The definition provided by UNFCCC (United Nations Framework Convention on Climate Change)defines as 'a change that is attributed directly or indirectly to human activity that alters the composition of the global atmosphere.

and that is in addition to natural climate variability observed over comparable time periods' (Government of India, 2013, p17).

Forests

In India, a shift towards wetter forest types in the north eastern region and drier forest types in the north western region is predicted.15 Current climate change scenarios predicting an increase of CO2 concentrations of approximately 575 parts per million (ppm) to 740 ppm are expected to result in large shifts in Indian forests, affecting up to 80% of the forest by the end of the century. More than half the vegetation is unlikely to adapt to the new conditions and will become susceptible to biotic and abiotic stress. Climate is an important determinant of vegetation patterns, and has significant influence on the structure and ecology of forests, meaning that specific plant communities are associated with particular climate zones.

Water

According to the 2006 Human Development Report of the UNDP, 2.5 billion people in South Asia will be affected by water scarcity by the year 205017. Rising temperature, changing precipitation patterns, and an increasing frequency of extreme weather events are expected to be the main reasons for reducing regional water availability and impacting hydrological cycles of evaporation and precipitation. This will drastically affect agriculture production in a region where over 60 percent of the agriculture is rain fed, such as in India.

Melting glaciers

Glaciers play an important role in climate and weather regulation. They are important for maintaining ecosystem integrity and in feeding rivers, thus providing water for agriculture. Apart from the monsoon rains, India relies heavily on the Himalayan Rivers for its irrigation. The hydrological characteristics of Himalayan watersheds have already undergone significant changes as a result of climate change and anthropogenic activities. This has resulted in increased variability in rainfall and surface water runoff, more frequent hydrological disasters, and the pollution of lakes.

Melting glaciers will impact certain areas more than others and in different ways. The increase in temperature resulting in more rainfall instead of snow is a direct consequence of Himalayan glacial melt. More rain and less snow fall increases the availability of fresh water in the short term but decreases water availability in the long run since moisture is not stored in frozen form. The retreat of Himalayan glaciers reduces water flow into the rivers. A reduced water flow from the melting glaciers changes the watershed distribution and the ecological parameters of rivers causing the temperature of river waters to rise.

Disruption of rainfall patterns

The disruption of established precipitation patterns negatively impacts Indian agriculture since agriculture systems have developed cropping patterns dependent on regional weather conditions. Across regions, precipitation patterns are changing with wet years becoming wetter and dry years become drier. The development of crops is also affected by increase in intra rain fall variability. This change could result in a greater number of heavy rainfall events a decrease in the overall number of rainy days, and longer gaps between rains, as well as increased rate of evaporate-transpiration.

This would disturb established cropping patterns. Heavy rainfall combined with a decrease in the total number of rainy days is occurring over a major part of India. Annual rainfall variability is increasing.. Estimates show that the rise in mean surface temperature will not only affect the post-monsoon and winter weather, but it is likely to lead to a 70% decline in summer rainfall. These changes are altering the seasons and changing the distribution of fauna and flora. They are also affecting the emergence and spread of pathogens which affect crop yields.

Flooding

Changes in the rainfall pattern leading to a more frequent of intense occurrence weather events will lead to an increased risk of flooding. In tropical areas extreme events will impact agriculture that is already vulnerable to floods and environmental hazards such as drought, cyclones, and storms.

Droughts

In certain vulnerable arid and semiarid regions, increased temperatures have already resulted in diminished precipitation. Notably, precipitation in Southern Asia and Western Africa has decreased by 7.5% between 1900 and 200525. Increased temperatures cause an intensification of the water cycle with more extreme variations in weather events and longer-lasting droughts. Furthermore, the expected temperature increase is likely to exacerbate drought conditions during sub-normal rainfall years. Large areas in Rajasthan, Andhra Pradesh, Gujarat, and Maharashtra and some areas of Karnataka, Orissa, Madhya Pradesh, Tamil Nadu, Bihar, West Bengal, and Uttar Pradesh are already experiencing recurrent drought with several regions currently experiencing water deficits. In dry land areas, marginal cropland could convert to range-land, and some crop-land and rangelands could no longer be suitable for food and fodder production. More frequent drought would necessitate greater multi-year reservoir storage capacity, in which India is currently deficient. Water conservation and management practices, as well as water storage will need urgent attention.

Monsoon

Monsoons are the lifeline of Indian agriculture so it is not surprising that the changes occurring in monsoon patterns are damaging crop yields. The timely arrival of the monsoon is of crucial importance to food production in the country and changing patterns in the monsoon are a threat to agriculture, food security, and the overall economy. The onset of the summer monsoon in India is getting delayed and disturbed. This affects crop cycles and cultivation in rainfed areas. Monsoon delays and failures inevitably lead to a reduction in agricultural output, thereby deepening food insecurity. Pre-monsoon rainfall disruption can be just as big problem however.

Soil

While it is natural to expect precipitation patterns to be impacted by climate change, soil processes are also heavily affected. This is because changes in temperature and precipitation influence water run-off and erosion, affecting soil, organic carbon and nitrogen content and salinity in the soil. This in turn has a major impact on the biodiversity of soil micro-organisms. These parameters are very relevant to soil fertility. Higher air temperatures will increase soil temperature and with it, increase the speed of organic matter decomposition and other soil processes that affect fertility. Experiments show that global warming will have the effect of reducing soil organic carbon by stimulating decomposition rates.26 At the same time increasing CO2 can also have the effect of increasing soil organic carbon through net primary production. According to researchers, a small increase in temperature in low carbon soil results in higher carbon dioxide emissions as compared with medium and high carbon soil. this phenomenon makes low carbon soil more vulnerable to warming.

Biodiversity

The fourth IPCC Report (2007) states that by the end of this century climate change will be the main cause of biodiversity loss. If there is an increase of the average global temperature by 1.5-2.5°C, then approximately twenty to thirty percent of known plants and animal species will be threatened by extinction30. Climate change will increase the pressure on land degradation and habitat loss, as well as genetic erosion which is already intensifying because of the growing uniformity in agricultural systems across the world. By mid-century, most species could lose half of their geographical range and size because of habitat fragmentation. According to the Food and Agricultural Organization (FAO), three-quarters of the global crop diversity is already lost. This is particularly problematic as the loss of genetic diversity, both in natural ecosystems and domesticated crops, is exacerbating the impact of climatic change.

Crop productivity

Most studies on the impact of climate change on agriculture come to the same conclusion that climate change will reduce crop yield in the tropical area. According to the IPCC, the next few decades of climate change are likely to bring benefits to higher latitudes through longer growing seasons, but in lower latitudes, even small amounts of warming will tend to decrease yields.

The regional inequality in food production resulting from climate change will have a very great implication for global food politics. Even without the challenge of climate, food security is an issue in the tropical areas considering that almost 800 million people in the developing world are already suffering from hunger 34. The effects of global warming could leave no room for manoeuvre since in many parts of India, and other developing countries, crops are already being cultivated near their maximum temperature tolerance. This is especially true in the dry land, non-irrigated areas, where vulnerabilities are high. In these regions even moderate warming of 1°C for wheat and maize and 2°C for rice will reduce yields significantly.

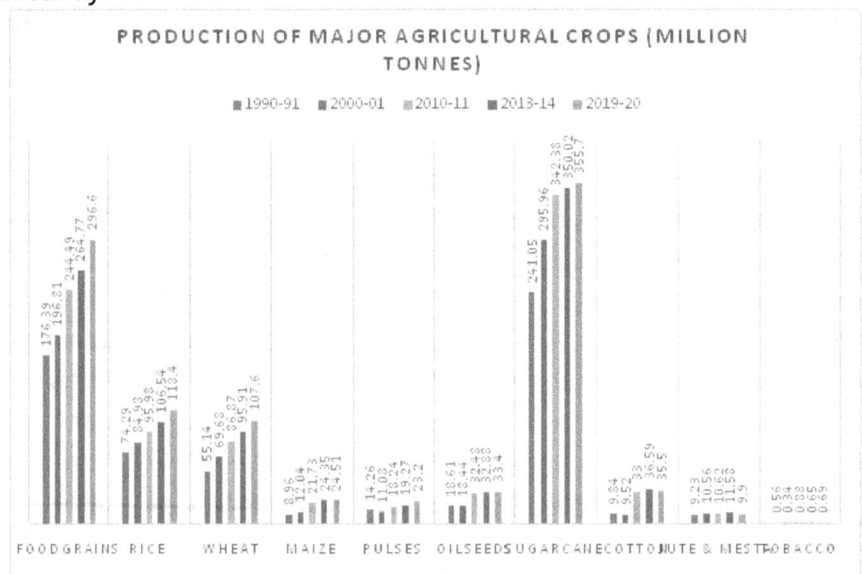

Production of major agricultural crops (Million tonnes)

Crops	1990-91	2000-01	2010-11	2013-14	2019-20
Food grains	176.39	196.81	244.49	264.77	296.6
Rice	74.29	84.98	95.98	106.54	118.4
Wheat	55.14	69.68	86.87	95.91	107.6
Maize	8.96	12.04	21.73	24.35	24.51

Crops	1990-91	2000-01	2010-11	2013-14	2019-20
Pulses	14.26	11.08	18.24	19.27	23.2
Oilseeds	18.61	18.44	32.48	32.88	33.4
Sugarcane	241.05	295.96	342.38	350.02	355.7
Cotton	9.84	9.52	33.00	36.59	35.5
Jute&Mesta	9.23	10.56	10.62	11.58	9.9
Tobacco	0.56	0.34	0.88	0.65	0.69

Source: Directorate of Economics and Statistics, Department of Agriculture and Cooperation.

Climate change and Indian agriculture

Agriculture lies at the heart of many fundamental global challenges faced by humanity including food security, economic development, environmental degradation, ecological balances and climate change. There is no humanitarian goal more crucial than feeding a world population projected to expand beyond nine billion by 2050. Meeting increases in food demands associated with growing population and income levels is likely to require increases in total food production of 50 percent or more by mid-century.

Agriculture provides employment for 2.6 billion people worldwide and accounts for 20 to 60 percent of the gross domestic product of many developing countries, forming the backbone of rural economies, contributing to local employment, and ensuring food security for poorer populations (California Environmental Associates, 2014, p11). Global climate change is a change in the long-term weather patterns that characterize the regions of the world. The term "weather" refers to the short-term (daily) changes in temperature, wind, and/or precipitation of a region (Merritts et al. 1998). In the long run, the climatic change could affect agriculture in several ways such as quantity and quality of crops in terms of productivity, growth rates, photosynthesis and transpiration rates, moisture availability etc.

Climate change is likely to directly impact food production across the globe. Increase in the mean seasonal temperature can reduce the duration of many crops and hence reduce the yield.

According to A K Singh, deputy director-general (natural resource management) of the Indian Council of Agricultural Research (ICAR), medium-term climate change predictions have projected the likely reduction in crop yields due to climate change at between 4.5 and 9 per cent by 2039.

The long run predictions paint a scarier picture with the crop yields anticipated to fall by 25 per or more by 2099. With 27.5% of the population still below the poverty line, reducing vulnerability to the impacts of climate change is essential. Indian food production must increase by 5 million metric tons per year to keep pace with population increase and ensure food security. Coping with the impact of climate change on agriculture will require careful management of resources like soil, water and biodiversity. To cope with the impacts of climate change on agriculture and food production, India will need to act at the global, regional, national and local levels.

Agriculture and allied activities contribute 24.9% (2000-01) to the gross domestic product (GDP), with agriculture (includes all crops, animal husbandry and dairying) contributing 22.7% and forestry and fishing at 1 and 1.2% respectively. The share of agriculture and allied activities has fallen from 57.7% in 1950-51, when agriculture contributed 50% of the GDP, with forestry contributing 6.7% while fishing contributed 0.89%. In absolute terms, contribution of agriculture and allied activities has more than tripled from about 81000 crores to 290000 crores (at 1993-94 prices).

Agriculture is the dominant land-use type with a net sown area of around 143 Mha since the 1970s, accounting for 47% of the reporting area. The forest areas occupy 22.5% of the area, while 3.6% is under permanent pasture and grazing lands, 4.7% is culturable wasteland and 7.6% is under current and other fallows. Around 34% of the area under agriculture is sown more than once, amounting to a gross cropped area of 191 Mha (1997/98).

The share of forests has increased from 14% to 22% of the reporting area though it still falls much shorter of the targeted 33% of the total landmass, according to the National Forest Policy. Net area sown has increased from 42% to 47% while, area sown more than once as a percentage of total cropped area has shot up from 10% to 24%. Fallow lands have fallen from 10 to 8% of the reported area, culturable waste lands have fallen from 8 to 4.5%, a huge decline has been notedin the area under land under miscellaneous tree crops and groves from 7 to 1% a slight increase from 2.3 to 3.6% in area under permanent pastures and other grazing lands, and a fall in the area not available for cultivation from 17% to 14%. Cereals account for about 53%, pulses account for 13%, thus the area under food grains occupies 65% of the total gross cropped area, commercial crops as a whole (including oilseeds, sugarcane, tea, coffee, cotton, jute, mesta, tobacco etc) make up 25% with oilseeds alone accounting for 15% and horticultural crops at 4%. Among the important crops, rice occupies 23% while wheat occupies 13% with jowar and bajra occupying 6 and 5% respectively among cereals. Sugarcane occupies 2.3% while groundnut occupies 4%, rapeseed and mustard 3.3%, cotton almost 5%. Over time, in terms of relative shares in cropped area, rice has more or less retained its share, while wheat occupied only about 7.6% of the area in 1950-51, jowar had a higher share at 11.8%, bajra at 7.4%. Total cereals as a whole have actually fallen from 61% to the current 53%. Pulses were at 15.6%. Sugarcane has risen from a mere 1.3% and horticulture crops were at 1.7%.

Total oilseeds share has gone up from 8.3% while cotton has slightly improved from 4.3% in 1950-51. 67 In terms of employment patterns in agriculture sector, population of total workers has increased from 140 to 402 million while the total population has risen from 361 million in 1951 to 1027 million in 2001. The share of rural population has declined from 82.7% to 72.22%, while that of cultivators in the workforce has declined from 49.9% to 31.7% and that of agricultural laborers has risen from 19.5% to 26.7% in the workforce. (Source: Indiastat.com).

More than 60% of the country's population still depends on the primary sector directly or indirectly. The production of food grains has increased from 51 million tonnes in 1950-51 to 212 million tonnes in 2001-2002. The yield of food grains has risen from 522 kg/ha in 1950-51 to 1739 kg/ha in 2001-02 while that of cereals has risen from 542 kg/ha to 1983 kg/ha in the same period.

The growth rate of agricultural production was 3.2% in the first plan period of 1951-56, dipped to a negative of 0.8% in the third plan 1961-66, peaked at 6.2% in the sixth plan period of 1980-85 and was at 4.1% in the eight plan, 1992-97. Crop-wise, while rice, wheat and oilseeds surged forward during the eighties, there was stagnation in the production of coarse cereals, pulses, cotton, jute and sugarcane (Nadkarni, 1993). Fish production rose from 752 million tonnes in 1950-51 to 5262 in 1998-99. In terms of geographical spread, the five states of Punjab, Haryana, Uttar Pradesh, Uttaranchal, Bihar and West Bengal contribute 50% of the total rice production in India, 82% of wheat production, 57% of total cereal production and 55% of total food grain production while the semiarid states of Andhra Pradesh, Karnataka, Tamil Nadu, Goa, Maharashtra, Gujarat, Orissa, Madhya Pradesh, Jharkhand and Chattisgarh contribute 42% of total rice production, 9% of wheat production, 33% of total cereal production, 62% of total pulses production and 35% of the total food grain production.

These states account for 52% of the total cropped area (56% of the net sown area), while the five states of the Indo Gangetic Plain account for 31% of net cropped area and 27% of the net sown area. The hilly states of north east and north west Himalayas account for 4% of the net cropped area and account for 4.25% of total food grain production. Rajasthan accounts for 11% of the area and 5% of the total food grain production. So there is necessary to conserve the environment for the purpose of agriculture as well as future.

Conclusion

India has made one of the fastest progress in the world, in addressing its environmental issues and improving its environmental quality.Still, India has a long way to go to reach environmental quality similar to those enjoyed in developed economies. Pollution remains a major challenge and opportunity for India.Environmental issues are one of the primary causes of disease, health issues and long term livelihood impact for India. Yajnavalkya Smriti, a historic Indian text on statecraft and jurisprudence, suggested to have been written before the 5th century AD, prohibited the cutting of trees and prescribed punishment for such acts. Kautalya's Arthashastra, written in Mauryan period, emphasised the need for forest administration. Ashoka went further, and his Pillar Edicts expressed his view about the welfare of environment and biodiversity. Coping with the impact of climate change on agriculture will require careful management of resources like soil, water and biodiversity. To cope with the impact of climate change on agriculture and food production, India will need to act at the global, regional, national and local levels.

References

- Van Der Warf, Hayo; Petit, Jean (December 2002). "Evaluation Of The Environmental Impact Of Agriculture At The Farm Level: A Comparison And Analysis Of 12 Indicator-Based Methods".Agriculture, Ecosystems and Environment 93 (1-3): 131–145. Doi:10.1016/S0167-8809(01)00354-1. Retrieved 21 April 2015.
- Jump Up^ "Un Report On Climate Change" (Pdf). Archived From The Original (Pdf) On 2007-11-14. Retrieved 25 June 2007.
- Jump Up^ Hance, Jeremy (May 15, 2008). "Tropical Deforestation Is 'One Of The Worst Crises Since We Came Out Of Our Caves'". Mongabay.Com / A Place Out Of Time: Tropical Rainforests And The Perils They Face.
- Government Of India (2013-14), Economic Survey 2013-14, Ministry Of Finance, New Delhi.

- Government Of India (2012-13), State Of Agriculture 2012-13, Ministry Of Agriculture, New Delhi
- Government Of India (2013), Statistics Related To Climate Change – India, Statistics & Programme Implementation Government Of India, New Delhi. Www.Mospi.Gov.In
- Prof. Dr. P. S. Kamble, Concerns Of Climate Change For Indian Agriculture.
- "India's National Communication To The United Nations's Framework Convention On Climate Change". 2004. Ministry Of Environment And Forest. Http://Unfccc.Int/Resource/Docs/Natc/Indnc1.Pdf Accesed On 01 February 2012.
- "Climate Change Impact In The Asia/Pacific Region." United Nations Convention To Combat Desertification. Http://Www.Ifad.Org/Events/Apr09/Impact/Pacific.Pdf Accessed On 01 February, 2012
- UNDP (2006). Beyond Scarcity: Power, Poverty, And Global Water Crisis. Human Development Report. UNDP
- "Water: Facts And Trends" In World Business Council For Sustainable Development Page.3. Http://Www.Unwater.Org/Downloads/Water_Facts_And_Trends.Pdf Accessed In 01 February 2012.
- Aggarwal.P.K. Edt, "Global Climate Change And Indian Agriculture, Case Studies From The Icar Network Project., Indian Council Of Agricultural Research, New Delhi.. Page No.Viii

Chapter 9

ISBN: 978-81-948672-7-2

Agricultural Exports Policy in India: A Critical Appraisal

DR.T.P.SHASHIKUMAR
Assistant Professor
Department of Economics
Karnataka State Open University , Mysore
Email: shashiksou85@gmail.com

Abstract: India, with a large and diverse agriculture, is among the world's leading producer of cereals, milk, sugar, fruits and vegetables, spices, eggs and seafood products. Indian agriculture continues to be the backbone of our society and it provides livelihood to nearly 50 per cent of our population. India is supporting 17.84 per cent of world's population, 15% of livestock population with merely 2.4 per cent of world's land and 4 per cent water resources. Hence, continuous innovation and efforts towards productivity, pre & post-harvest management, processing and value-addition, use of technology and infrastructure creation is an imperative for Indian agriculture. Therefore, agro processing and agricultural exports are a key area and it is a matter of satisfaction that India's role in global export of agricultural products is steadily increasing. India is currently ranked tenth amongst the major exporters globally as per WTO trade data. India's share in global exports of agriculture products has increased from 1% a few years ago, to 2.2 %.The Agricultural Export Policy was introduced to increase and provide support to productivity, pre and post-harvest management, value-addition and upgrade technology. The decline of agricultural commodities reduced GDP and other economic affairs. To make major reforms to the export policy for agriculture, India restructured from the Green Revolution Era and promoted Agricultural Export Policy to diversify the food and non-food agriculture base to emerge as a leading player in the world in agricultural trade. The policy would increase the agricultural exports leading to stable growth in GDP, benefits for farmers, employment in rural areas, quality and scope for value addition and future market potential.

Keywords: Agriculture, Development, Export Policy, Productivity, Investment, Farmers, Technology

Introduction

Agriculture is described as the backbone of Indian economy mainly because of three reasons. One, agriculture constitutes largest share of country's national income though the share has declined from 55 percent in early 1950s to about 25 percent by the turn of the Century. Two, more than half of India's workforce is employed in its agriculture sector. Three, growth of other sectors and overall economy depends on performance of agriculture to a considerable extent. Besides, agriculture is a source of livelihood and food security for large majority of vast population of India. Agriculture has special significance for low income, poor and vulnerable sections of rural society. Because of these reasons agriculture is at the core of socio economic development and progress of Indian society, and proper policy for agriculture sector is crucial to improve living standards and to improve welfare of masses.

Agricultural policy describes a set of laws relating to domestic agriculture and imports of foreign agricultural products. Governments usually implement agricultural policies with the goal of achieving a specific outcome in the domestic agricultural product markets. Agricultural policies use predetermined goals, objectives and pathways set by an individual or government for the purpose of achieving a specified outcome, for the benefit of the individual(s), society and the nations' economy at large. Agricultural policies take into consideration the primary, secondary and tertiary processes in agricultural production. Outcomes can involve, for example, a guaranteed supply level, price stability, product quality, product selection, land use or employment.

The Agricultural Export Policy was introduced to increase and provide support to productivity, pre and post-harvest management, value-addition and upgrade technology. The policy was introduced after the stagnant agricultural trade in the international market from 2013-2017. The decline of agricultural commodities reduced GDP and other economic affairs. To make major reforms to the export policy for agriculture, India restructured from the Green Revolution Era and promoted Agricultural Export Policy to diversify the food and non-food agriculture base to emerge as a leading player in the world in agriculture trade. The policy would increase the agricultural exports leading to stable growth in GDP, benefits for farmers, employment in rural areas, quality and scope for value addition and future market potential.

The Agricultural Export Policy shall address the following areas to increase the exports in agriculture:

- Changes in the food patterns
- Increasing unstable incomes
- Shrinking farming area
- Changing socio-economic, agro-climatic and dietary patterns
- Transportation and Infrastructure
- Post-harvest losses
- Generating employment
- Minimising the loss in the value chain

Vision of the Policy

To utilise the potential of Indian Agriculture and emerge as a global power in agriculture through a suitable framework and policies.

Objective of the policy

The important objective of the agriculture export policy in India are as follows,

- To create a stable trade policy regime.
- To double the agricultural exports to 100 Billion USD by 2022.
- Diversify the export base and increase the high value and value-added agricultural exports.
- Promote ethnic, traditional, organic and non-traditional agri products.
- Create a platform to provide a mechanism for better market access to deal with sanitary and phytosanitary issues.
- To increase India's share in the world agri exports by integrating with international value chain.
- Create a framework to support farmers to increase benefits and opportunities in the overseas market.

National Agriculture policy 2000: An overview

The broad objectives of NAP is to actualize the vast untapped growth potential of Indian agriculture; strengthen rural infrastructure to support faster agricultural development; promote value addition; accelerate the growth of agro business; create employment in rural areas; secure a fair standard of living for the agricultural population; discourage migration to urban areas; and face the challenges arising out of economic liberalization and globalization.

Over the next two decades, NAP aims to attain a growth rate in excess of four percent per annum in the agriculture sector that is based on efficient use of resources and conserves our soil, water and bio-diversity and growth that is demand driven and caters to domestic markets and maximises benefits from exports of agricultural products in the face of the challenges arising from economic liberalization and globalization. The NAP has indicated a package of policy initiatives to achieve the objectives. These include Sustainable Agriculture, Food and Nutritional Security, Generation and Transfer of Technology, Inputs Management, Incentives for Agriculture, Investments in Agriculture, Institutional Structure, Risk Management and Management Reforms.

Being the first ever-released agriculture policy by the government, NAP sets out the intensions of strengthening the base of the most important sector of Indian economy in the following way:
• It aims to attain a growth rate in excess of four percent per annum in the agriculture sector and seeks to achieve this growth in a sustainable manner and with equity;
• It talks about investment in agriculture in terms of capital formation and stepping up the public investments for narrowing regional imbalances;
• It emphasises the need to increase the need for the exports of agricultural products;
• It talks about institutional and management reforms for the sector and about managing the risks in the sector; and
• It emphasises easy availability of credit and other inputs to the farming community.

In addition, the policy addresses the enhanced level of efficiency of input use consistent with environmental upholding and makes clear the vision of the government to fortify this sector.

Agriculture export policy 2018: An overview

India, with a large and diverse agriculture, is among the world's leading producer of cereals, milk, sugar, fruits and vegetables, spices, eggs and seafood products. Indian agriculture continues to be the backbone of our society and it provides livelihood to nearly 50 per cent of our population. India is supporting 17.84 per cent of world's population, 15% of livestock population with merely 2.4 per cent of world's land and 4 per cent water resources. Hence, continuous innovation and efforts towards productivity, pre & post-harvest management, processing and value-addition, use of technology and infrastructure creation is an imperative for Indian agriculture. Various studies on fresh fruits and vegetables, fisheries in India have indicated a loss percentage ranging from about 8% to 18% on account of poor post-harvest management, absence of cold chain and processing facilities. Therefore, agro processing and agricultural exports are a key area and it is a matter of satisfaction that India's role in global export of agricultural products is steadily increasing. India is currently ranked tenth amongst the major exporters globally as per WTO trade data for 2016. India's share in global exports of agriculture products has increased from 1% a few years ago, to 2.2 % in 2016.

Vision and objectives

The Agriculture Export Policy is framed with a focus on agriculture export oriented production, export promotion, better farmer realization and synchronization within policies and programmes of Government of India.

Vision

Harness export potential of Indian agriculture, through suitable policy instruments, to make India global power in agriculture and raise farmers income.

Objectives

• To double agricultural exports from present ~US$ 30+ Billion to ~US$ 60+ Billion by 2022 and reach US$ 100 Billion in the next few years thereafter, with a stable trade policy regime.

• To diversify our export basket, destinations and boost high value and value added agricultural exports including focus on perishables.

• To promote novel, indigenous, organic, ethnic, traditional and non-traditional Agri products exports.

• To provide an institutional mechanism for pursuing market access, tackling barriers and deal with sanitary and phytosanitary issues.

• To strive to double India's share in world agri exports by integrating with global value chain at the earliest.

• Enable farmers to get benefit of export opportunities in overseas market.

Policy recommendation of agriculture export policy 2018

The policy recommendations in this report are organized in two broad categories - strategic and operational.

Strategic

1. Policy Measures - Discussions with public and private stakeholders across the agricultural value chain highlighted certain structural changes that were required to boost agricultural exports. These comprise of both general and commodity specific measures that may be urgently taken and at little to no financial cost. The subsequent gains, however, are aplenty.

2. Infrastructure and logistics - Presence of robust infrastructure is critical component of a strong agricultural value chain. This involves pre-harvest and post-harvest handling facilities, storage & distribution, processing facilities, roads and world class exit point infrastructure at ports facilitating swift trade. Mega Food Parks, state-of-the-art testing laboratories and Integrated Cold Chains are the fundamentals on which India can increase its agricultural exports. Given the perishable nature and stringent import standards for most of the food products, efficient and time-sensitive handling is extremely vital to agricultural commodities.

3. Holistic approach to boost exports - Agricultural exports is determined by supply side factors, food security, processing facilities, infrastructure bottlenecks and several regulations. This involves multiple ministries and state departments. Strategic and operational

operational synergy across ministries will be key to boosting productivity and quality.

4. Greater involvement of State Governments in Agriculture Exports.

Critical appraisal of agricultural policy in India

Its criticisms of the Agriculture Export Policy include that:

• The document does not stress the need for improved irrigation as a precondition to higher growth

• Creating the required magnitude of irrigation would require trebling of public investment in real terms

• The NAP proposes to put India's 79.5 million hectares of wasteland to use for agriculture and afforestation, but does not elaborate any strategy to do so; most of this land requires heavy capital investments in order to make it productive

• The importance and implications of the increasing strain on India's limited water resources are not adequately recognized by the NAP

• NAP is silent on important institutional issues such as participatory management of irrigation, water, forest and common lands.

Policy recommendation and suggestions

The paper specific recommendations for future agricultural policy include:

- Care is needed to ensure that enough common property wasteland remains around inhabited areas to serve the needs of resource-poor rural communities who live there
- Awareness about the value of water and its sustainable use needs to be raised
- Water conservation measures such as rainwater harvesting should be put in place
- Panchayati Raj institutions should be given guidelines on taking care of healthy breeding practices for livestock
- Technology policy needs to ensure both that appropriate technologies are generated and that they are effectively disseminated to end users
- Interactions of Indian National Agricultural Research System scientists with science institutions in the west need to be strengthened.

- Competition should be promoted in the seed sector by encouraging large scale private sector participation in the seeds business area specific agricultural development strategies should be devised, taking into account groundwater status, soil health and other characteristics.

Major constrains to the agricultural export policy in India
General observation

There are several reasons that have proved as a constraint in the successful implementation of the NAP though it intends to reinforce the agriculture sector but lacks at some serious fronts. It is said to harm the interests of Indian farmers in certain areas, including:

• It came at a time when there existed clear and disturbing signs of a declining trend in food grain productivity, fast emerging barriers to sustainability of agriculture, depleting underground water resources, the ever-growing indebtedness in farming and farmers suicides. It has been criticised on the ground that inspite of the prevalent situation the policy demanded that the farmers should diversify and produce export oriented crops and the focus shifted from agriculture to industry, trade and commerce.

• It is felt that it has set unrealistic targets. No clear-cut strategy has been evolved or mentioned for achieving the respective targets. It seems like other policies wherein the plans are evolved but implementation is too far-fetched.

• It looks more like a document citing issues involved in agriculture than any serious statement of policy and intent by the Central Government. The document though contains a set of policy intentions and explains exhaustively what has to be done but ignores to include the procedure for the implementation of the set intentions.

• It does not explain as to how the implementation shall be done and how the goals and objectives shall be achieved. It does not have any time bound and concrete action plan to carry out the activities mentioned in the document. No strategy has been formulated to implement the agenda mentioned in the document. It does not mention any deadline or time frame for the accomplishment of any task and lacks the target based result oriented approach.

• It was introduced in the year 2000 and since then no document has been released to gauge the status of the intentions of the government mentioned in the document. There has never been any supplement to NAP document since then and no follow up mechanism has been evolved.

• Above all, conflict is about agriculture being the state subject while policy is formed by the Central Government, which plays an important role in the recommendation of national agricultural policies. There are pronounced disparities among the states in agricultural progress. Some states are agriculturally much progressed and some states lag behind. A centralised policy tends to create constraints as it fails to cater to the state specific needs, albeit the states find it difficult to work in accordance with the national policy. This makes NAP quite incongruent, especially when the formulation of such a policy is made with little or no participation by State Governments.

Conclusion and Recommendations

The policy should aim at providing productive employment to the huge labour force dependent on agriculture sector. One way of doing this could be by making the sector competitive enough to provide gainful employment to all. But as the labour force seems to increase at a much higher rate than land productivity, efforts should be made to direct the excess labour to nonfarm activities as it would help them earn better giving rise to the income of those left in agriculture and this process should be clearly defined by the policy.

The Central Government should involve varied stakeholders in the process of policymaking and implementation and should try to strengthen the existing mechanism of this sector. The efforts of the centre by establishing National Commission on farmers should be further strengthened. The actual condition of the farmers and the ground realities they face have to be kept in mind while formulating any policy as they get affected the most. The priorities should be set up in accordance with the farmers needs. Schemes introduced by the government to improve the condition of the farmers should focus on effective implementation with regular rounds of checks and follow-ups. Also the policies formulated by government should be time bound and sector specific. Each target should be followed by a strategy.

action plan. Proper planning and strategy should be in place to assess what percent of growth has to be achieved in a specified timeframe. A white paper needs to be published after a specific interval of time and a follow up committee should evaluate the policy in the light of facts mentioned in the white paper and policy should be altered accordingly. The states being different from each other in terms of agriculture growth, all agricultural policies including price supports, input subsidies, produce marketing, and consumer subsidies must ultimately be agreed to, and implemented, by the State Governments.

Importance of the state participation in the formulation of the policy should be realized. Also, the State Government should reconcile with the policy made by the centre and support it in the implementation process. Operational co-ordination between centre and state team is very crucial for successful implementation of the policy. It is very true that the NAP has to be compatible with the imperatives of economic liberalization and globalization but safeguarding the interests of the farmers should be the priority of the government. Agriculture education and research can be important tools in making the farmers compatible with the changing environment. The NAP needs to be brought to the centre stage urgently. It must be examined with a sense of urgency. The current state of all the resources, on which the country's agriculture is based, must be revisited thoroughly.

References

- Clarkson Nancy and Kishore G. Kulkarni, (2011). "Effects of India's Trade Policy on Rice Production and Exports", working paper,pp:01-23.
- Kalal.D.Shoyabahmed (2010). "Competitiveness of Indian Agricultural Exports: Pattern and Policy" Southern Economist.vol.49,no.12.pp:45-48.
- Khan Amir Ullah (2011)."Indian Agriculture and the WTO: A developing country perspective"working paper, no.121,pp:65-87.
- LallSanjaya (2004). "Selective Policies for Export Promotion", Research for Action 43,pp:12-62.
- Pandey B.N. and Mishra(2008). "Impact of Indian Agricultural Trade",Trade Review,vol.6(2),pp:177-190.
- Tantri L. Malini and Deshpande R.S. (2010) "WTO and Agricultural Policy in Karnataka", working paper 259,pp:04-25.

Chapter 10   **ISBN: 978-81-948672-7-2**

Organic Farming as an Opportunity for Sustainable Agriculture Development in India

RAJESHA. B

Assistant Professor
Department Of Economics
Government First Grade College, Kyathanahalli,
PandavapuraTq. Mandya District
Email :rajeshab02@gmail.com

Abstract: Agriculture, and hence choices made by a farmer, has a significant impact on environmental sustainability. Presently, more than 98% of farmers in India follow conventional farming using chemical fertilizers, often to the detriment of the environment, yields, and personal health. They remain hesitant to adopt organic farming despite its promise of greater sustainability and profitability.

These approaches ultimately disturb the nutrient balance of the soil and therefore reduce soil fertility . To deal with the existing problem, Organic farming provides a natural way of crop cultivation by using environment friendly, animal and plant-based local organic resources that are highly enriched in nutrients required for crop plants. It enhances microbial activities and increases soil health.

Organic farming is an efficient and promising agricultural approach for environmental sustainability as it provides yield stability, improved soil health, no environmental concerns, organic food, and a reduction in the use of synthesized fertilizers. There are different agricultural approaches working to reduce environmental concerns but the use of organic farming, is no doubt, the best scientifically proven environment-friendly approach to maintaining an ecological balance of our agriculture and ecological systems.

This paper provides an overview of organic farming opportunities for sustainable agriculture development in India.

Keywords: Organic Forming, Sustainable Agriculture, Fertility of Soil and nutrition food

Introduction

The Concept of Organic Farming Organic agriculture is a safe. Sustainable farming system producing healthy crops without damage to the environment. It avoids the use of artificial chemical fertilizers and pesticides on the land, relying instead on developing a healthy, fertile soil and growing a mixture of crops. Sustaining agricultural productivity depends on quality and availability of natural resources like soil and water. Agricultural growth can be sustained by promoting conservation and sustainable use of these scarce natural resources through appropriate location specific measures. Indian agriculture remains predominantly rainfed covering about 60% of the country's net sown area and accounts for 40% of the total food production. Thus, conservation of natural resources in conjunction with development of rain fed agriculture holds the key to meet burgeoning demands for food grain in the country. Towards this end, National Mission for Sustainable Agriculture (NMSA) has been formulated for enhancing agricultural productivity especially in rain fed areas focusing on integrated farming, water use efficiency, soil health management and synergizing resource conservation.

Objectives

- To maintain and enhance long-term fertility of soils.
- To produce healthy, nutritious and quality food.

Methodology of the Study

The study is mainly descriptive in nature. Secondary data are used for the purpose of the study. Secondary data was collected from websites, various articles and journals.

Limitation of the Study

Lack of primary data: Time consuming. As the research mainly depends on secondary data, it may not be hundred percent accurate. The study is restricted to India only.

Conceptual framework
Organic Farming

The term "organic farming" was coined by Lord Northbound in 1940. "Organic agriculture is a holistic production management system which promotes and enhances agro-ecosystem health, including biodiversity, biological cycles, and soil biological activity. It emphasises the use of management practices in preference to the use of off-farm inputs, taking into account that regional conditions require locally adapted systems. This is accomplished by using, where possible, agronomic, biological, and mechanical methods, as opposed to using synthetic materials, to fulfill any specific function within the system." (FAO/WHO Codex Alimentarius Commission, 1999). Organic farming can be defined as an agricultural process that uses biological fertilisers and pest control acquired from animal or plant waste.

Benefits of organic farming

Economical: In organic farming, no expensive fertilisers, pesticides, or HYV seeds are required for the plantation of crops. Therefore, there is no extra expense.

Good return on Investment: With the usage of cheaper and local inputs, a farmer can make a good return on investment.

High demand: There is a huge demand for organic products in India and across the globe, which generates more income through export.

Nutritional: As compared to chemical and fertilizer utilized products, organic products are more nutritional, tasty, and good for health.

Environment-friendly: The farming of organic products is free of chemicals and fertilisers, so it does not harm the environment.

Challenges of organic forming in India lack of Awareness

The most important constraint felt in the progress of organic farming is the inability of the government policy making level to take a firm decision to promote organic agriculture.

Output marketing problems

It is found that before the beginning of the cultivation of organic crops, their market ability and that too at a premium over the conventional produce has to be assured. Inability to obtain a premium price, at least during the period required to achieve the productivity levels of the conventional crop will be a setback.

Shortage of bio-mass

Many experts and well informed farmers are not sure whether all the nutrients with the required quantities can be made available by the organic materials. Even if this problem can be surmounted, they are of the view that the available organic matter is not simply enough to meet the requirements.

Inadequate supporting infrastructure

In spite of the adoption of the NPOP during 2000, the state governments are yet to formulate policies and a credible mechanism to implement them. There are only four agencies for accreditation and their expertise is limited to fruits and vegetables, tea, coffee and spices. The certifying agencies are inadequate.

High input costs

The small and marginal farmers in India have been practicing a sort of organic farming in the form of the traditional farming system. They use local or own farm renewable resources and carry on the agricultural practices in an ecologically friendly environment. However, now the costs of the organic inputs are higher than those of industrially produced chemical fertilizers and pesticides including other inputs used in the conventional farming system.

Marketing problems of organic inputs

Bio-fertilizers and bio-pesticides are yet to become popular in the country. There is a lack of marketing and distribution network for them because the retailers are not interested to deal in these products, as the demand is low. The erratic supplies and the low level of awareness of the cultivators also add to the problem. Higher margins of profit for chemical fertilizers and pesticides for retailing, heavy advertisement campaigns by the manufacturers and dealers are other major problems affecting the markets for organic inputs in India.

Low yield

In many cases the farmers experience some loss in yields on discarding synthetic inputs on conversion of their farming method from conventional to organic. Restoration of full biological activity in terms of growth of beneficial insect populations, nitrogen fixation from legumes, pest suppression and fertility problems will take some time and the reduction in the yield rates is the result in the interregnum. It may also be possible that it will take years to make organic production possible on the farm.

High MRP

It is almost obvious that due to the extreme care taken to go along with organic farming, the results would be kept at a high price.

Opportunities of organic forming in India

Promotion of natural farming under Bharatiya Prakritik Krishi Padhati (BPKP) of PKVY has been initiated to encourage use of natural on-farm inputs for chemical free farming. Andhra Pradesh and Kerala have taken up one lakh hectare and 0.8 lakh hectare area respectively for promotion of natural farming under BPKP. Similarly, continuous area certification and support for individual farmers for certification have also been initiated during 2020-21 to bring in default organic areas and willing individual farmers under the fold of organic farming. State agencies, Primary Agricultural Credit Societies (PACS), Farmer Producer Organisations (FPOs), entrepreneurs among others can avail loans for setting up of post-harvest infrastructure for value addition to organic produce under 1 lakh crore Agriculture Infrastructure Fund (AIF) of Aatmanirbhar Bharat.

(The story is based on the information given by the Union Minister of Agriculture and Farmers Welfare, Narendra Singh Tomar in the Rajya Sabha today — September 18,2020). The Government of India provides assistance for promoting organic farming across the country though different schemes.

1. Param paragat krishi vikasyojana (PKVY)

Param paragat Krishi Vikas Yojana promotes cluster based organic farming with PGS (Participatory Guarantee System) certification. Cluster formation, training, certification and marketing are supported under the scheme. Assistance of Rs. 50,000 per ha/3 years is provided out of which 62 percent (Rs. 31,000) is given as incentive to a farmer towards organic inputs.

2. Mission organic value chain development for north eastern region (MOVCDNER)

The scheme promotes third party certified organic farming of niche crops of north east region through Farmer Producer Organisations (FPOs) with focus on exports. Farmers are given assistance of Rs 25,000 per hectare for three years for organic inputs including organic manure and bio-fertilisers among other inputs. Support for formation of FPOs, capacity building, post-harvest infrastructure up to Rs 2 crore are also provided in the scheme.

3. Capital investment subsidy scheme (CISS) under soil health management scheme

Under this scheme, 100 percent assistance is provided to state government, government agencies for setting up of mechanised fruit and vegetable market waste, agro waste compost production unit up to a maximum limit of Rs 190 lakh per unit (3000 Total Per Annum TPA capacity). Similarly, for individuals and private agencies assistance up to 33 percent of cost limit to Rs 63 lakh per unit as capital investment is provided.

4. National mission on oilseeds and oil palm (NMOOP)

Under the Mission, financial assistance at 50 percent subsidy to the tune of Rs. 300 per hectare is being provided for different components including bio-fertilisers, supply of Rhizobium culture, Phosphate Solubilising Bacteria (PSB), Zinc Solubilising Bacteria (ZSB), Azatobacter, Mycorrhiza and vermi compost.

5. National food security mission (NFSM)

Under NFSM, financial assistance is provided for promotion of biofertiliser (*Rhizobium*/PSB) at 50 percent of the cost limited to Rs 300 per hectare.The Indian government has been promoting safer agricultural practices and the Union Budget 2022 will focus on organic farming and natural farming. The government will also stress on domestic oilseed production to reduce edible oil imports, said Finance Minister Nirmala Sitharaman in her Budget 2022 speech.n the Union Budget 2022-23, some key announcements have been made for the agriculture sector and the farmers. "The year 2023 has been announced as the year of millets," announced Finance Minister Nirmala Sitharaman while presenting the Union Budget 2022-23. Finance minister went on to inform that chemical-free natural farming will be promoted throughout the country, with focus on farmers' lands in 5-km-wide corridors along the river Ganga in the first stage.   To reduce dependence on import of oil seeds, government will implement a rationalized and comprehensive scheme to increase domestic production of oil seeds, she said.

Current status of organic farming in India

Organic farming is recognized as a sustainable agriculture practice promoting use of organic / bio inputs that takes care of environment (including soil, biodiversity) and human wellbeing. Organic farming practices such as crop rotations, inter-cropping, cover crops, use of organic fertilizers and minimum tillage improves biodiversity, nutrient content and water retention capacity of the soil. Organic agriculture also contributes to mitigating the greenhouse effect and global warming through its ability to sequester carbon in the soil (FAO).

Cultivable land area under organic farming has increased from 11.83 lakh hectare in 2014 to 29.17 lakh hectare in 2020 due to the focused efforts of the government. Over the years, the organic promotion activities led to development of state specific organic brands, increased domestic supply and exports of organic produce from north eastern region. Taking cue from the success of the organic initiatives, a target of 20 lakh hectare additional area coverage by 2024 is envisaged in the vision document.   Awareness programmes, availability of adequate post- harvest infrastructure, marketing facilities, premium price for the organic produce among others would certainly motivate farmers towards organic farming thereby increasing organic coverage in the country. During the year 2020-21, 9.86 lakh farmers have been brought under Organic Farming. Under the schemes PKVY & MOVCDNER, abput 9.48 lakh ha area in India is under organic farming as per the Ministry of Agriculture & Farmers' Welfare.

Organic foods are safe and healthy and also boost immunity. With increased awareness on organic foods, people are inclining more towards organic products and it is fact that since the advent of COVID-19, the demand of organic product has increased in domestic market. The domestic market is growing @ 17% and the projected demand of organic food market is likely to cross Rs. 87.1 crore by 2021 from the Rs. 53.3 crore in 2016 (ASSOCHAM–EY). The export market size of organic products has increased by 42% during the current year (2021) compared to last year and the volume of organic products exported from April 2020 to February 2021 is 819250 MT with value of 948 million USD. In view of increasing demand of organic food and realizing the advantage of chemical free farming for environment and human life, Government of India has been promoting Organic farming in the country through dedicated schemes. Namely Paramparagat Krishi Vikas Yojana (PKVY) and Mission Organic Value Chain Development for North Eastern Region (MOVCDNER) since 2015-16. Both the schemes stress on end to end support to organic farmers i.e. from production to certification and marketing.

Sr. No	State	Position	Area under organic certification (in million ha)
1	China	3rd	3.14
2	USA	7th	2.02
3	India	9th	1.94
4	Brazil	12th	1.18

Comparative data in regard to organic cultivation with other countries (Source: NAPCC)

NMSA derives its mandate from Sustainable Agriculture Mission which is one of the eight Missions outlined under National Action Plan on Climate Change (NAPCC). The strategies and programmers of actions (POA) outlined in the Mission Document, that was accorded 'in principle' approval by Prime Minister's Council on Climate Change (PMCCC) on 23.09.2010, aim at promoting sustainable agriculture through a series of adaptation measures focusing on ten key dimensions encompassing Indian agriculture namely; 'Improved crop seeds, livestock and fish cultures', 'Water Use Efficiency', 'Pest Management', 'Improved Farm Practices', 'Nutrient Management', 'Agricultural insurance', 'Credit support', 'Markets', 'Access to Information' and 'Livelihood diversification'. During XII Five Year Plan, these measures are being embedded and mainstreamed onto ongoing/proposed Missions/ programmes / Schemes of Dept. of Agriculture & Cooperation (DAC&FW) through a process of restructuring and convergence. NMSA architecture has been designed by converging, consolidating and subsuming all ongoing as well as newly proposed activities/programmes related to sustainable agriculture with a special emphasis on soil & water conservation, water use efficiency, soil health management and rainfed area development. The focus of NMSA will be to infuse the judicious utilization of resources of commons through community based approach.

Conclusion

Organic farming yields more nutritious and safe food. The popularity of organic food is growing dramatically as consumer seeks the organic foods that are thought to be healthier and safer. Thus, organic food perhaps ensures food safety from farm to plate. The organic farming process is more eco-friendly than conventional farming. Organic farming keeps soil healthy and maintains environment integrity thereby, promoting the health of consumers. Moreover, the organic produce market is now the fastest growing market all over the world including India. Organic agriculture promotes the health of consumers of a nation, the ecological health of a nation, and the economic growth of a nation by income generation holistically.

References

• Altieri, M. A. 2000. The ecological impacts of transgenic crops on agro ecosystem health. Ecosystem Health, 6 (1):13-23

• Brown, L. (2004). Outgrowing the earth, the food security challenge in an age of falling water tables and rising temperatures. New York: W._W. Norton.

• Haas, G., F. Wetterich, and U. Kopke. 2001. Comparing intensive, extensified and organic grassland farming in southern Germany by process life cycle assessment. Agriculture, Ecosystems and Environment, 83:43-53.

• Hansen, B., H. Fjelsted, and E. S. Kristensen. 2001. Approaches to assess the environmental impact of organic farming with particular regard to Denmark. Agriculture, Ecosystems and Environment, 83:11-26.

• Liebhardt, W. C. 2001. Get the facts straight: organic agriculture yields are good. Organic Farming Research Foundation Information Bulletin, Summer (No. 10): 1-7.

• Rigby, D. and D. Caceres. 2001. Organic farming and the sustainability of agricultural systems. Agricultural Systems, 68:21-40.

Web References

• https://academic.oup.com › fqs › article

•https://www.en.krishakjagat.org/national-news/status-of-organic-farming-in-india/

• https://en.gaonconnection.com/union-budget-2022

 http://employmentnews.gov.in/newemp/MoreContentNew.aspx

•https://blog.agthentic.com/organic-farming-in-india-challenges-and-opportunities

• https://agriculturepost.com/farm

• https://agricoop.nic.in

Chapter 11

ISBN: 978-81-948672-7-2

A Study of Sustainable Development and Its Impact on Rural Development in India.

Dr. Kiran, S. P
Assistant Professor
H. K. Veeranna Gowda,Degree College,
Madduru, Mandya District
Email: kirueco@gmail.com

Dr. Manjuprasad, C.
Faculty Member
Department of Studies in Economics and
Cooperation, University of Mysore, Mysuru

Abstract: Rural development as an old theme has been evolving new strategies and fresh contents. The earlier definition of rural development assumed the village to be a homogenous entity. Rural Poverty in rural areas though poverty is hidden, it is real and predominant than urban areas. Poverty of course is a tragedy, but dimensions of urban poverty are documented well and even glamorized in music and film industries. But the rural poor are still overlooked and considered backward. Sustainable Rural Development is improving the quality of life for the rural poor by developing capacities that promote community participation, health and education, food security, environmental protection and sustainable economic growth, thereby enabling community members to leave the cycle of poverty and achieve their full potential. This paper contains the main objectives is study the goals of rural development, objectives of rural development, rural poverty and sustainable development goals in India.

Keywords: Rural, Poverty, Health, Education, Development, Sustainable.

Introduction

Rural development has been one of the most formidable and fundamental aspect of India's developmental efforts. The concept of development since seventies has undergone a change and has become more comprehensive. There are different concepts of rural development which have been propounded by various policy makers In the post independence period.

Rural development as an old theme has been evolving new strategies and fresh contents. The earlier definition of rural development assumed the village to be a homogenous entity. But, this myth was exploded under the impact of first step of programmes undertaken for rural development. Rural development, in-fact concentrated on the poverty eradication programmes to uplift the weaker sections of the community in the backward areas. Rural development has remained a priority item through successive five year plans. It is a concept whose contents have been of great variations. The concept of rural development took birth in the context of agricultural development and it remained for a long time conterminous with agriculture improvement in India.

Objectives and Methodology of the study

The main objectives of the study to understand the goals of rural development, objectives of rural development, rural poverty and sustainable development goals in India and the study is based on only secondary sources, the secondary sources collected from published articles, books, Journals, News Paper and website sources.

Goals of Rural Development

Since rural development is a preferred condition, the approaches and strategies of rural development adopted in various countries depend on the ideological orientating of the life of the people and the structure of national needs. The policies and programmes for rural development is a strategy for the improvement of socio-economic and political life of the people with special emphasis on the rural poor. In a democratic society, there are three goals of rural development.

1. Raising community solidarity
2. Raising agricultural needs and
3. Institutionalization of equality

Rural Poverty

Rural Poverty in rural areas though poverty is hidden, it is real and predominant than urban areas. Poverty of course is a tragedy, but dimensions of urban poverty are documented well and even glamorized in music and film industries.

But the rural poor are still overlooked and considered backward. Fighting against rural poverty is multidimensional. To live a better life, each individual's core desire is to survive, sustain, and prosper thereby making changes to live fearless and hunger less. Leaving aside desires and aspirations yet one fifth of the world's population is devoid of fulfilling basic physiological needs. Besides the usual complaints in the countryside – access to services (hospitals, education, work, transportation, and even patchy mobile phone coverage and broadband) and vulnerability to wide range of diseases – it is the seemingly invisible nature of poverty that is necessary to be tackled in order to combat rural poverty.

Rural poverty needs to be targeted and ended to achieve many of the Sustainable Development Goals but most particularly goal 1: No Poverty. According to the International Union for Conservation of Nature's article entitled "Sustainable Development, Poverty and The Environment: A Challenge to the Global Community," development can be sustainable only when economic well-being, social development, and environmental stability are addressed together. Sustainable Development Goals (SDGs), the redefined, refocused Millennium Development Goals (MDGs), are comprehensive wherein both social development and environmental stability areaddressed in all its 17 Goals. SDGs were carefully framed so that no individual is excluded. As some regions had growth in par with globalization and urbanization, yet in some areas it led to resource deprivation particularly in rural areas (Schuftan 2003).

Sustainable Development Goals (SDGs)

The Sustainable Development goals (SDGs) are successors to the 'Millennium Development Goals MDGs'. The MDGs were adopted in 2000 by governments to make global progress on poverty, education, health, hunger and the environment. The MDGs expired at the end of 2015.   During 25-27 September 2015, the member states of the United Nations converged in New York for the United Nations (UN) Summit on Sustainable Development and adopted the new global goals for sustainable development.

The world leaders pledged their commitment to the new '2030 Agenda for Sustainable Development', encompassing 17 universal and transformative SDGs. The United Nations General Assembly has taken the resolution, for adopting 17 Sustainable Development Goals (SDGs), with 169 targets and 304 indicators, on 25th September, 2015 under the official agenda "Transforming our world : the 2030 Agenda for Sustainable Development".

The SDGs are a universal set of goals, targets and indicators that all UN member states are expected to use to frame their development agendas, socio-economic policies, and actions towards low carbon pathways for the next 15 years, in order to achieve a sustainable world where 'no one is left behind' without compromising sustainability of the planet. These new global goals are much broader and comprehensive than the outgoing MDGs, as they attempt to address all three dimensions of sustainable development- economic, social and environmental.

Societal and environmental threats in rural areas

Rural communities are facing several challenges in the context of climate change, land degradation, deforestation, biodiversity loss, and fragmentation of natural habitats, poverty, and geographical isolation. The rural population is more prone to extreme poverty, famine, social exclusion, and environmental injustice, particularly in developing countries from Africa, Asia, and Latin America. Rural communities depend on local geographical conditions (climate, natural resources, landscape, and geographical barriers, socioeconomic conditions, demographic features) to develop agricultural, industrial, or tourism activities as economic development pathways.

A traditional economy based on subsistence agriculture is still widespread across rural regions of the globe. This type of economy is volatile to natural hazards (extreme weather, flash floods, landslides, erosion, drought) and poor agricultural productivity which translates into famine, extreme poverty, land abandonment, and massive migration.

Land use management is a key factor for future rural development perspectives and to find the optimal equilibrium between natural habitats, agricultural lands, and built-up areas. It reveals the emerging societal and environmental threats, sectoral approaches, and synergic effects that must be addressed at subnational levels by each country via regional and local authorities towards rural areas.

Rural areas must cope with social, demographic, economic, governance, and environmental challenges. As an example, extensive cattle ranches and emerging oil palm cultivation threaten biodiversity conservation and food security across tropical rural regions while increasing social inequalities and conflicts. On the other hand, agricultural land abandonment (associated with traditional farming, low productivity, poor infrastructure, aging population, massive migration, land ownership change, political instability) has created several socioeconomic and ecological dysfunctional ties in southeastern Europe.

Poor agricultural productivity in the Global South is related to the low use of improved seed, use of inappropriate fertilizer, inadequate irrigation, and lack of incentives for farmers in the absence of remunerative markets. Extreme poverty, hunger, and undernourishment and rural depopulation are critical issues to be solved across rural Africa besides the poor access to critical amenities. Climate changes, land fragmentation, natural resource depletion, political instability, corruption, and conflict areas will further threaten rural areas of developing countries.

In this context, rural resilience and circular economy are key strategic directions to further develop rural economies and reduce socioeconomic inequalities and environmental injustice coupled with access to proper education. A linear economy based on "take make-dispose" model feed by consumerism society is harmful for the environment and long-term sustainability of urban and rural areas.

Strategies for Poverty Eradication in Rural Ares through Panchayati Raj Institution and also Convergence with the other line Departments.

- Water conservation and water harvesting to help poor farmers in increasing agriculture production.
- Micro and minor irrigation for increasing production and productivity in agriculture and horticulture sector.
- Renovation of traditional water bodied to enhance fish production by poor fish farmers.
- Tree plantation and horticulture for increasing fruit production and enhance the income of the rural horticulturist.
- Land development / waste land development for enhancing production and productivity among small and marginal farmers.

Sustainable Rural Development

Sustainable Rural Development is improving the quality of life for the rural poor by developing capacities that promote community participation, health and education, food security, environmental protection and sustainable economic growth, thereby enabling community members to leave the cycle of poverty and achieve their full potential. This can be achieved by documenting lessons-learned in the field on practical solutions to challenges facing the rural poor. Sharing these lessons in a systematic framework so that under served villagers receive the critical information on development activities they need to begin improving their difficult lives. The Process The dynamics of development processes reveal that rural development is a sequence of four, clearly identifiable development phases.

Key Themes Found in Sustainable Rural Development are
- Community Participation.
- Water and Sanitation.
- Health and Hygiene.
- Conservation and the protection of natural resources and the environment.
- Poverty Reduction & Disease Education.
- Food Security and Agriculture.
- Greater survival prospects for mothers and their infants.
- Education & Equal Opportunities for Women.

- Economic Growth and Infrastructure Development
- Science & Technology

Government Schemes for Rural Entrepreneurship in India

- Prime Minister Employment Generation Programme.
- Centrally Sponsored Schemes (CSS) of Export Market Promotion.
- Dairy Entrepreneurship Development Scheme (DEDS).
- Development of Khadi, Village and Coir Industries.
- Pm Mudra Yojana - MSME Loan / SME Loan Scheme.
- Entrepreneurship Development Institution Scheme.

Conclusion

This paper concluded that the Rural development has been one of the most formidable and fundamental aspect of India's developmental efforts. Rural development has remained a priority item through successive five year plans. Rural development took birth in the context of agricultural development and it remained for a long time conterminous with agriculture improvement in India. The policies and programmes for rural development is a strategy for the improvement of socio-economic and political life of the people with special emphasis on the rural poor. Rural poverty needs to be targeted and ended to achieve many of the Sustainable Development Goals but most particularly goal. Poverty and The Environment: A Challenge to the Global Community," development can be sustainable only when economic well-being, social development, and environmental stability are addressed together.

References

- Ana Nieto Masot and José Luis GurríaGascón. (2021). Sustainable Rural Development: Strategies, Good Practices and Opportunities. Land 2021, 10, 366.Https://Www.Mdpi.Com/Journal/Land.

- NeeruSinghal and Yuvika Singh.(2016). Sustainable Rural Development - Initiatives Taken by Government of India.International Journal of Management and Social Science Research Review, Vol.1, Issue.3.March- 2016 Page 207.

- Schuftan C (2003) Poverty and inequity in the era of globalization: our need to change and to re-conceptualize. Int J Equity Health 2(1):4. https://doi.org/ 10.1186/1475-9276-2-4.

- Silpa Immanuel VoolaAnd Prince Immanuel Kalyanasundaram. (2020). Rural Poverty And Sustainable Development Goal. Springer Nature Switzerland Ag 2020.Https://Doi.Org/10.1007/978-3-319-69625-6_48-1.

- Vikas Kumar. (2021). Rural Development Startegies and Approaches in India – A New Dimention. International journal of Multidisciplinary educational research, ISSN:2277-7881; Impact Factor :6.514(2021); IC Value:5.16; ISI Value:2.286, Peer Reviewed And Refereed Journal: Volume:10, Issue:3(1), March:2021.

Chapter 12

ISBN: 978-81-948672-7-2

The Role of Cooperatives in Sustainable Agricultural Development

DR. MANJUPRASAD, C.

Faculty Member
Department of Studies in Economics and Cooperation
Manasagangotri, University of Mysore
Mysuru – 570006
Email: prasad3083@gmail.com

Abstract: Sustainable agriculture is ecologically sound, economically viable, socially just and human. Cooperatives can contribute to all SDGs, both because they are involved in the very diverse economic sectors concerned, and because their impact contributes substantially to the global objectives pursued. Cooperatives are well-placed to contribute to sustainable development's triple bottom line of economic, social and environmental objectives plus the governance agenda, not least because they are enterprises that endeavour to meet the economic progress of members while satisfying their socio cultural interests and protecting the environment. The major agricultural cooperatives include production and marketing cooperatives, poultry and livestock cooperatives, fishing and fish marketing cooperatives, and food processing and marketing cooperatives. This study was undertaken to analyze Sustainable development, the cooperative values and principles and The Role of Cooperatives in Sustainable Agricultural Development.

Keywords: Sustainable, Agriculture, Cooperatives, Development, Marketing

Introduction

Sustainable development means, development at present meets the needs of the present generation without compromising the ability of future generation to meet their own demand. Sustainability in agriculture means the land and resources that use for agriculture today should be handed over to the future generations in a sustainable form so that they can continue to practice agriculture and have food security.

This means that we have to use lands, water resources, etc in such a manner that the future generations are also will be able to have sustainable development. Sustainable agriculture is the system of raising crops for greater human utility through utilization of resources with better efficiency without disturbing im balancing or polluting the environment. Sustainable agriculture is ecologically sound, economically viable, socially just and human.

The sustainable development goals

The sustainable development goals were negotiated in light of the missed targets of their immediate predecessor, the Millennium Development Goals (MDGs). They are unique in their commitment to all countries contributing toward their achievement, rather than just low to middle income nations. The SDGs, through a large number of indicators, cover all components of economic activity (agriculture, industry, housing, health, education, production, consumption etc.), and address a wide range of key global concerns (poverty, equality, employment, gender, climate change, peace etc.). Cooperatives can contribute to all SDGs, both because they are involved in the very diverse economic sectors concerned, and because their impact contributes substantially to the global objectives pursued.

Of course, a number of SDGs and their indicators may be more particularly well suited to the cooperative identity. Cooperatives are well-placed to contribute to sustainable development's triple bottom line of economic, social and environmental objectives plus the governance agenda, not least because they are enterprises that endeavour to meet the economic progress of members while satisfying their socio cultural interests and protecting the environment.

They offer an alternative model for social enterprise, with contributions to sustainable development well beyond job creation. Since cooperatives' share in GDP and total enterprises is currently relatively small in most countries, their promotion and expansion could be an important instrument for achieving the Sustainable Development Goals (SDGs).

The cooperative values and principles: hardwired for sustainable development

The first clear relation is at the very root of the definition provided by the International Co-operative Alliance, which defines a cooperative as 'an autonomous association of persons united voluntarily to meet their common economic, social, and cultural needs and aspirations through a jointly-owned and democratically-controlled enterprise.' Cooperatives meet a diversifiedset of needs, which go beyond profit generation or shareholder return. The wide range of needs identified, whilst not specifically clarifying an environmental aim in this case, continue to acknowledge that people, in order to voluntarily achieve well-being, require more than simple economic well-being. The emphasis on the commonalities suggests that one's need does not necessarily lead to the detriment of another, and links strongly to the cooperative value of solidarity.

The cooperative movement has indeed considered the role of environmental protection as having an implicit recognition within its values and principles. The most recent addition to the cooperative principles, 'Concern for community', was adopted at the Manchester Congress in September 1995, and included strong debate over the links between the cooperative movement and environmental protection. The principle reads; "While focusing on member needs, cooperatives work for the sustainable development of their communities..." Cooperatives, therefore, have a tangible relation to the communities within which they are based. Not only do cooperatives arise from a genuine need, when compared with the frequent manufacturing of consumer needs by conventional capitalist companies, but profits stay within and are reinvested by the community.

There are incentives therefore, both economic and social, to ensure this investment, in order for a community to come together to meet their needs through the formation of democratically accountable, member-based organisations.

The role of cooperatives in sustainable agricultural development

An agricultural cooperative is also called as a farmers' co-op. It is a cooperative where farmers pool their resources in certain areas of activity. It allows little farms to do what big farms can do, like buy inputs at bulk rates, increase volume to open new markets and lower the per-use cost of equipment.

Major agricultural cooperatives include production and marketing cooperatives, poultry and livestock cooperatives, fishing and fish marketing cooperatives, and food processing and marketing cooperatives. The major contributions of Cooperatives in Sustainable Agricultural Development as follows;

- Owing to the "one member – one vote" principle cooperatives provide an ideal organizational environment for the joint, equitable and democratic ownership and management of resources, and the provision of basic services.
- Cooperatives can play a key role in poverty reduction. While savings and credit cooperatives facilitate their member's access to financial capital, agricultural cooperatives help farmers access the inputs require to grow crops and keep livestock and help them process, transport and market their products.
- Cooperatives enhance (agricultural) productivity by generating economies of scale and scope through the joint use of modern and/or expensive equipment, the division of labour between members, joint pre- and post-production services such as input supply and output marketing, and the exchange of knowledge and innovation.
- Information sharing on prices and markets is a core function of rural cooperatives, in particular of unions and federations.
- Cooperatives can make specific contributions to employment generation through (a) worker-ownership; (b) joint labour contracting; (c) facilitation of access to resources, markets, land and finance; and (d) workers' takeover of bankrupt private or public enterprises.
- Cooperatives are contributing towards gender equality by expanding women's opportunities to participate in local economies.

- Cooperatives can act as transmission belts to foster the elimination of child labour and forced labour, in particular in agriculture.
- Cooperatives can contribute through a variety of means to sustainable tourism, and promote a fairer sharing of profits generated by tourism: guides' and porters' coops, handicraft coops, transport coops, fishery coops, etc.
- Cooperatives facilitate the sharing of commercial, financial and technological services among their members (including small-scale enterprises), through joint marketing, joint finance, joint processing, or other joint services. Through horizontal cooperation and vertical integration they enable their members to "climb up" the value chain, including through "fair trade".
- Financial cooperatives can create national and international networks to facilitate remittances; because of the principles of "people over profits" and "cooperation among cooperatives" they will seek to minimize the remittance fees.
- As per their principle of "concern for community" cooperatives provide an appropriate organizational framework for the joint management and efficient use of natural resources, including through recycling and reuse.
- Cooperatives enable artisanal fishers (both marine and in-land) with a variety of services that make their activity more sustainable and profitable: joint ownership of boats, joint purchasing of equipment, joint marketing, joint processing, etc.
- Cooperatives are institutions themselves; as per their values, principles and governance structures they are more transparent and accountable than other forms of private or public businesses and organizations. This also applies to higher-level cooperative structures, provided those are built according to cooperative principles.
- Cooperatives play a leading role in boosting agricultural exports because they can organize export-related services that are out of reach to individual producers. By doing so they may encourage small-scale farmers to grow export crops, and thus increase overall exports of the country.

Conclusion

Sustainability in agriculture means the land and resources that use for agriculture today should be handed over to the future generations in a sustainable form so that they can continue to practice agriculture and have food security. Cooperatives can contribute to all SDGs, both because they are involved in the very diverse economic sectors concerned, and because their impact contributes substantially to the global objectives pursued. Cooperatives can play a key role in poverty reduction. While savings and credit cooperatives facilitate their member's access to financial capital, agricultural cooperatives help farmers access the inputs require to grow crops and keep livestock and help them process, transport and market their products and cooperatives have an important role to play in ensuring inclusive sustainable development.

References

- Ahmet Candemir And Sabine Duvaleix. (2021). Agricultural Cooperatives and Farm Sustainability – A Literature Review. Journal of Economic Surveys (2021) Vol. 35, No. 4, Pp. 1118–1144.
- Alok K. Sahoo, Sanat K. Meher, Tarak C. Panda, Susrita Sahu, Rukeiya Begum and N. C. Barik (2020).Critical Review on Cooperative Societies in Agricultural Development in India.Current Journal of Applied Science and Technology, https://www.researchgate.net/publication/343635885, 39(22): 114-121, 2020; Article no.CJAST.5993, ISSN: 2457-1024.
- Anantha Selvam S. (2015). Sustainable Development in Indian Agriculture. Shanlax International Journal of Economics, Vol. 4, No. 1, December 2015, ISSN: 2319-961X.
- Arundhati Kulkarni. (2013). Sustainable Development: Roles Of Societies Governments And NGOS. International Journal of Environmental Science: Development And Monitoring (Ijesdm) Issn No. 2231-1289, Volume 4 No. 2 (2013).
- Basil Hans V. An Analysis of Sustainable Agricultural Development in India. Electronic Copy Available At: Https://Ssrn.Com/Abstract=3150442

- Cooperatives and the Sustainable Development Goals. A Contribution to The Post-2015 Development Debate a Policy Brief. International Labour Office, Geneva. International Cooperative Alliance.
- Cooperatives and the Sustainable Development Goals:The role of cooperative organisations in facilitating SDG implementation at global, National and Local levels. International Cooperative Alliance.
- Cynthia Giagnocavo, Emilio Galdeano-Gómez and Juan Carlos Pérez-Mesa. (2018). Cooperative Longevity and Sustainable Development in a Family Farming System. Sustainability, 27th June 2018, 10, 2198; Doi:10.3390/Su10072198 Www.Mdpi.Com/Journal/Sustainability.
- Esther Gicheru. The Role of The Co-operative Enterprise Model In Implementing The Sustainable Development Goals (SDGS) In Least Developed Countries (LDCS).The Cooperative University of Kenya, Nairobi, Kenya.
- Francisca Castilla-Polo and M. Isabel Sánchez-Hernández. (2020). Cooperatives and Sustainable Development: A Multilevel Approach Based on Intangible Assets. Sustainability, 17th May 2020, 12, 4099; Doi:10.3390/Su12104099 Www.Mdpi.Com/Journal/Sustainability
- Homiga U. (2014). Sustainable Development and the Role of Cooperatives. The Cooperator, Vol 52, No 6, Pp 25-32, December 2014.
- Jurgen Schwettmann. The Role of Cooperatives in Achieving the Sustainable Development Goals. International Labour Office, Geneva.
- Krishnachetty Ravichandran. (2020). PACS and SDGs: A Study in Tamil Nadu, India. International Journal of Co-operative Accounting and Management, Volume 3, Issue 2, 2020.
- Michael Gertler (2001). Rural Co-operatives and Sustainable Development. Centre for the Study of Co-operatives University of Saskatchewan, ISBN 0–88880–443–1.
- Yashoda. (2017). Role of Primary Agricultural Co-Operative Society (PACS) in Agricultural Development in India. Global Journal of Management and Business Research: C Finance. Volume 17, Issue 3, Version 1.0, 2017. Online ISSN: 2249-4588 & Print ISSN: 0975-5853.

Chapter 13

ISBN: 978-81-948672-7-2

A Study on Organic Farming in India

DR. J. L. BANASHANKARI

Assistant Professor
Department of Economics
Karnataka State Open University,
Mukthagangothri, Mysuru
Email: banashreeraj@gmail.com

Abstract: In recent years, organic farming is gradually becoming popular because of its added merit. Products produced in organic and natural farming are getting demand both in national and international markets. Organic farming is considered as one of the widely used methods which is again recognized as the best alternative for avoiding ill effects of chemical farming done with the use of chemical fertilisers and pesticides.Organic farming system in India is not new and is being followed from ancient time. It is a method of farming system which primarily aimed at cultivating the land and raising crops in such a way, as to keep the soil alive and in good health by use of organic wastes (crop, animal and farming wastes) and other biological materials along with beneficial microbes (bio-fertilisers) to release nutrients to crops for increased sustainable production in an eco-friendly pollution free environment.Present paper focuses on present status of organic farming in India and also discussed the benefits, problems of organic farming in India.

Keywords: Organic Farming, Fertilisers, Pesticides, Eco-Friendly, Agriculture.

Introduction

In recent years, organic farming is gradually becoming popular because of its added merit. Products produced in organic and natural farming are getting demand both in national and international markets. Organic farming is considered as one of the widely used methods which is again recognized as the best alternative for avoiding ill effects of chemical farming done with the use of chemical fertilisers and pesticides.

Organic farming system in India is not new and is being followed from ancient time. It is a method of farming system which primarily aimed at cultivating the land and raising crops in such a way, as to keep the soil alive and in good health by use of organic wastes (crop, animal and farming wastes) and other biological materials along with beneficial microbes (bio-fertilisers) to release nutrients to crops for increased sustainable production in an eco-friendly pollution free environment.

Definition

As per the definition of the United States Department of Agriculture (USDA) study team on organic farming "organic farming is a system which avoids or largely excludes the use of synthetic inputs (such as fertilisers, pesticides, hormones, feed additives etc) and to the maximum extent feasible rely upon crop rotations, crop residues, animal manures off-farm organic waste, mineral grade rock additives and biological system of nutrient mobilization and plant protection".

FAO suggested that "organic agriculture is a unique production management system which promotes and enhances agro-ecosystem health, including biodiversity, biological cycles and soil biological activity and this is accomplished by using on-farm organic agronomic, biological and mechanical methods in exclusion of all synthetic off-farm inputs". As defined by International Organic Agricultural Movement, "Organic agriculture is a production system that sustains the health of soil, eco-systems and people. It relies on ecological processes, biodiversity and cycles adapted to local conditions, rather than the use of inputs with adverse effects.

Organic agriculture combines tradition, innovation and science to benefit the shared environment and promote fair relationships and a good quality of life for all involved". Thus organic farming is a kind of natural farming where all organic inputs are applied on agriculture to reap maximum benefit of natural farming without having any adverse effect on agricultural crops, on land, on farmers or on environment.

As the people throughout the world are gradually aware of the adverse effect of agricultural crops raised with the help of chemical fertilisers and pesticides on human health thus, the demand for organic farming products are increasing day by day both from within and outside the country. The prices of organic farm products are comparatively higher in the market due to its increasing cost conditions. In spite of that, demand for organic crops is increasing day by day.

History of Organic Farming

The concept of organic agriculture were developed in the early 1900s by Sir Albert Howard, F. H. King, Rudolf Steiner, and others who believed that the use of animal manures (often made into compost, cover crops, crop rotation, and biologically based pest controls resulted in a better farming system. Howard, having worked in India as an agricultural researcher, gained much inspiration from the traditional and sustainable farming practices he encounted there and advocated for their adoption in the west. Such practices were further promoted by various advocates such as J. I. Rodale and his son Robert, in the 1940s and on ward, who published organic gardening and farming magazine and a number of texts on organic farming. The demand for organic food was stimulated in the 1960s by the publication of silent spring, by Rachel Carson, which documented the extent of environmental damage caused by insecticides.

Organic food sales increased steadily from the late 20th century. Greater environmental awareness, coupled with concerns over the health impacts of pesticide residues, and consumption of genetically modified (GMO) crops, fostered the growth of the organic sector.

In the United States retail sales increased from $ 20.39 billion in 2008 to $ 47.9 billion in 2019, while sales in Europe reached more than $ 52 billion in 2019.The price of organic food is generally higher than of conventionally grown food. Depending on the product, the season, and the vagaries of supply and demand, the price of organic food can be anywhere10 percent below to more than 100 percent above that of conventionally grown produce.

Need of Organic Farming

With the increase in population our compulsion would be not only to stabilize agricultural production but to increase it further in sustainable manner. The scientists have realized that the green revolution with high input use has reached a plateau and is now sustained with diminishing return of falling dividends. Thus, a natural balance needs to be maintained at all cost for existence of life and property. The obvious choice for that would be more relevant in the present era, when these agrochemicals which are produced from fossil fuel and are not renewable and are diminishing in availability. It may also cost heavily on our foreign exchange in future.

Organic Farming in India

Organic farming in a nascent stage in India. About 2.78 million hectare of farmland was under organic cultivation as of March 2020, according to the Union Ministry of Agriculture and farmers' welfare. This is two percent of the 140.1 million ha net sown area in the country. A few states have taken the lead in improving organic farming coverage, as a major part of this area is concentrated only in a handful of states. Madhya Pradesh tops the list with 0.76 million ha of area under. Organic cultivation that is over 27% of India's total organic cultivation area. The top three states-Madhya Pradesh, Rajasthan and Maharashtra- account for about half the area under organic cultivation. The top 10 states account for about 80% of the total area under organic cultivation.

Sikkim is the only Indian state to have become fully organic so far. A majority of the states have only a small part of their net sown area under organic farming.

Even the top three states that account for the largest area under organic cultivation- Madhya Pradesh, Rajasthan and Maharashtra- have only around 4.9, 2.0 and 1.6 percent of their net sown area under organic farming respectively. A few states such as Meghalaya, Mizoram, Uttarkhand, Goa and Sikkim have 10% or more of their net sown area under organic cultivation. All these states, except Goa, are in hilly regions.

Union territories such as Delhi, Dadra and Nagar Haveli and Daman and Diu, Lakshadweep and Chandigarh also have 10% or more of their net sown area under organic cultivation, but their agricultural area is very small. Almost all other states have less than 10% of their net sown area under organic. Currently, only around 12 states-Madhya Pradesh, Gujarat, Telengana, Sikkim, Bihar, Karnataka, Odisha, Rajasthan, UttarKhand, Chhattisgarh, Tamil Nadu and Uttar Pradesh-have their own state organic certification agencies accredited by Agricultural and Processed Food Products Export Development Authority (APEDA). Some states have either developed or are still in the process of farming organic brands such as MP organic, Organic Rajasthan, Nasik Organic, Bastar Naturals, Kerala Naturals, Jaivik Jharkhand, Naga Organic, Organic Arunachal, Organic Manipur, Tripura Organic and Five Rivers by Punjab.

Uttaranchal and some other state governments have already declared their states as "organic" state and created Special Export Zones like Basmati Export Zone. A large area of North Eastern States and other states may be developed commodity based "organic" production areas with greater political will and investment in research extension and marketing infrastructure more of this potential could be realized.

In order to attain increasing sustainability in agriculture the government of India recently introduced the National Mission for Sustainable Agriculture (NMSA) in order to popularize organic and natural farming.

The main objective of NMSA is to make agriculture more productive, sustainable, remunerative and climate resilient by promoting location specific integrated/composite farming systems and to conserve natural resources through application of appropriate soil and moisture conservation measures. Moreover, the Government has been promoting organic farming in the country through the scheme such as Paramparagat Krishi Vikas Yojana (PKVY) and Rashtriya Krishi Vikas Yojana (RKVY).

In the revised guideline for PKVY scheme issued during in the 2018, various organic farming models like Natural Farming, Vedic Farming, Cow Farming, Homa Farming, Zero Budget Natural Farming (ZBNF) etc. have been included wherein flexibility lies with the states to adopt any model of organic farming depending on the farmer's choice. Under the RKVY scheme, organic farming/ natural farming project components are considered by the respective State Level Sanctioning Committee (SLBC) according to their own priority/choice.

Zero Budget Natural Farming (ZBNF) has its unique character. The main aim of ZBNF is elimination of chemical pesticides and promotion of good agronomic practices. ZBNF also aims to sustain agricultural production with eco-friendly processes aligning with nature for producing agricultural produce free of chemicals. Here, soil fertility and soil organic matter is restored by pursuing ZBNF. Under ZBNF, less volume of water is required and it is a climate friendly agriculture system. In India, this programme is now being implemented in 131 clusters covering 704 villages under RKVY and 1,300 clusters covering 268 villages under PKVY. So far 1,63,034 farmers are practicing ZBNF. Organic farming is also being promoted through the scheme Mission Organic Value Chain Development for North Eastern Region (MOVCDNER) under National Mission for Agriculture (NMSA).In the meantime, some degree of progress has been attained in this direction. Some of the states which are progressively practicing ZBNF are Karnataka, Himachal Pradesh and Andhra Pradesh. After the introduction ZBNF, Andhra Pradesh has witnessed a sharp decline in input costs and also improvement in yields.

In India, the organic sector is growing by almost 17 percent every year and it has the potential to grow at a faster rate on rising demand for health and wellness food products, across the world. So, in times to come, the present trend is likely to grow much faster. As per the estimates of Ministry of Agriculture, the organic food market will touch ₹ 75,000 crore in the next five years (2020-25). Moreover, food processing value addition and organic food would be instrumental in realizing the government's target to double farmer's income.

Benefits of Organic Farming

Organic farming is a kind of agricultural practice which maintains harmonious relationship with nature to the maximum level without creating any bad impact or threat to the environment. Various countries of the world, especially the development and developing countries, are trying to attain higher growth in agricultural production by applying chemical fertilizers, pesticides leading to spreading contamination of poisonous chemicals in human food, feed, fodder and fibre. All these have been resulting serious health hazards for all categories of population. Under this scenario, organic farming can serve as a remedy to these problems. Organic farming has same benefits of its own. Following are some of the benefits of organic.

Farming

The most important benefit of organic farming is it provides healthy foods to all. Organic farming results improvement in soil quality. Maintaining proper soil quality is considered as the foundation on which organic farming is based. It has also been observed that proper adoption of organic farming can raise crop productivity and also the income of farmers. Organic farming may also lead to low incidence of pests. As for example, the effectiveness of organic cotton cultivation on pests is much higher as compared to conventional farming. Organic farming results increasing employment opportunities as it requires more labour input than the conventional farming. Practice of organic farming result several indirect benefits to farmers as well as consumers. While the farmers adopted organic farming are indirectly benefited from improvement in soil quality and productivity in farm production and environment.

Problems of Organic Farming

Although organic farming has its edge over conventional farming but it is subjected to some basic problems or constraints. One of the important problem faced organic farming is the lack of awareness of the farmers on the use of bio-fertilisers and bio-pesticides. Organic farming is also subjected to problem of output marketing of organic crops at a premium price over traditional crops. Under organic farming, the availability of organic matter is not simple enough to meet the requirement.

The small and marginal farmers practicing organic farming are facing the problem of high input costs. Supporting infrastructure required for organic farming is inadequate in India. There is lack of marketing and distribution network for organic inputs like bio-fertilisers and bio-pesticides as their demand is still low. India is well known in the world organic market for its organic or herbal products. Initiatives are taken for the production and export of organic products but the supply of organic products are not available regularly to meet such huge demand for its exports. Although, there a large number of brands of organic manures but there is lack of quality and standards for these bio-manures. Promotion of organic agriculture both for domestic consumption and exports needs appropriate agriculture policy support from the government. But policy support available till now is either absent or inappropriate. In India, political and social factors are favouring traditional farming with the objectives of dispensing favours for electoral benefits. Providing subsidy or supplying inputs like power, water either free of cost or at subsidized rate are taken to achieve political interest. Unfortunately such political and social factors usually block the encouragement spirit for developing organic farming.

Exports of Organic Food Products

Demand for Indian organic food products is increasing steadily in recent years. In 2018-19, India exported organic products worth ₹ 5,151 crore as compared to that of ₹ 3,453 crore in the previous year.In 2018-19, India produced around 2.67 million ton of certified organic products which include all varieties of food products, nearly oil seeds, sugar cane, cereals and millets, cotton, pulses, tea, fruits, spices, dry fruits, vegetables, coffee, medicinal plants. The production of organic products is not limited to the edible sector but also include produces like organic cotton fibre and functional food products. The Agricultural and Processed Food Products Export Development Authority (APDA) is a statutory body under the Commerce Ministry which is looking after the exports of organic products. Major demands under the organic product category are for flax seeds, sesame and soyabean; pulses such as arahar (red gram) Channa (pigeon pea);and rice along with tea and medicinal plants.

There is a growing demand for organic food from Canada, Taiwan and South Korea in recent years. Germany is one of the biggest importers of Indian organic products.

Now many new countries are also taking interest of Indian organic products. Demand for organic agricultural products is constantly increasing worldwide as organic products are grown without the use of chemical fertilisers and pesticides. As on March 31, 2019, total area under organic certification process was 3.56 million hectare. This includes 1.94 million hectare cultivable area and another 1,49 million hectare for wild harvest cultivation.

Among all the states, Madhya Pradesh has covered the largest area under organic certification followed by Rajasthan, Maharashtra and Uttar Pradesh. During 2016, Sikkim had achieved a remarkable distinction of converted its entire cultivation land (more than 76,000 hectare) under organic certification.

The total volume of export during 2018-19 was 6.14 lakh M.T. The organic food export realization was around ₹ 5,151 crore ($ 757.49 million). Organic products are exported to USA. Europeon Union, Canada, Switzerland, Australia, Isreal, South Korea, Vietnam, New Zealand and Japan.In order to boost the demand of organic agri products further both in India and abroad, APEDA, has been showcasing India's strengthas major hub for organic food products, ingredients, commodities and processed food in trade fairs in the name of 'Bio-fach India'.

Suggestions for Improvement

In India, conventional farming is subjected to many ill effects, which are related to unsustainability of agricultural production, health and sanitation problems, environmental problems etc. On the other hand, organic farming is gaining its momentum as an alternative method of farming. In the meantime, many countries have been able to convert around 2 to 10 percent of their cultivated land into organic farming. However, some areas need immediate attention in order to spread organic farming throughout the country.

Following are some suggestions in this regard

1. Substantial financial support is provided by both central and state governments in order to promote organic farming.
2. In order to increase sales of organic products both within and outside the country, steps be taken for market development organic products.
3. In order to maintain proper quality and standard of organic products, the producer organization should be encouraged to get itself accredited for, inspection and certification in accordance with the NSOP.
4. Arrangement is made for vigorous campaign so as to highlight the benefits of organic farming for raising awareness of both farmers and consumers.
5. Arrangement is made for identification of crops for cultivation on organic farms.
6. Steps are taken by the government and corporate bodies for export promotion of organic products produced by organic farmers of our country.

Conclusion

In recent years, awareness of the harmful effect of chemical-based fertilisers and pesticides on our health is on a rise. Conventional agriculture relies heavily on chemical fertilisers and toxic pesticides etc., which enter the food supply, penetrate the water sources, harm the livestock, deplete the soil and devastate natural eco-system. Efforts in evolving technologies which are eco-friendly are essential for sustainable development and one such technology which is eco-friendly is organic farming.Organic food is growing in popularity across the world. Many countries have around 10 percent of their food system under organic farming. There are many retail chains and supermarkets which are accorded with green status to sell organic food. Moreover, organic foods command higher price of around 10-1000 percent than conventional ones. Studies across countries have shown that organically grown food has more nutritional value than chemical farming thus providing us with healthy foods. Since organic farming requires more labour input than conventional farming, India will find organic farming an attractive proposition.

Finally, the produce is pesticide-free and produced in an environmentally sustainable way.

References:

1. P. K. Dhar, (2021), "Indian Economy-Its Growing Dimensions", Kalyani Publishers, Ludhiana.
2. Indian Economic Development (2021), National Council of Educational Research and Training, Karnataka Text Book Society, Bangaluru.
3. Shir Y. V. Singh and J. P. S. Dabas in Kurukshetra Journal, Vikas Pedia. In
4. Raoul Adamchak, Organic Farming Agriculture, btitanica.com
5. TNAU AGRITECH PORTAL, Organic Farming,
6. Agritech.tnau.ac.in/org
7. Brainy.in
8. Topperlearning.com
9. On a tardy trail: state of organic farming in India, downtoearth.org.in

Chapter 14

ISBN: 978-81-948672-7-2

Government Efforts in Promoting Sustainable Agriculture in India: A Study

CHANDRA PRASAD HOSAMANI
Research Scholar
Department of Economics,
Karnataka State Open University,
 Mysore
Email: shashiksou85@gmail.com

DR. SHASHI KUMAR. T.P.
Assistant Professor
Department of Economics,
Karnataka State Open University,
 Mysore

Abstract: Agriculture is a basic necessity for survival of human being. Post Green Revolution Indian agriculture is facing new challenges. The Government of India at Central and various state governments. are engaged in addressing the challenges posed due to various factors. The present research paper wants to analyze the schemes which are framed towards development of sustainable agriculture in India. The present research paper wants to analyze the schemes which are framed towards development of sustainable agriculture in India. The Government of India has brought many programs, policies and plans in order to meet the sustainable development goals envisioned by the United Nations, especially goals related to the goal of ending hunger achieving food security and improving nutrition and to promote sustainable agriculture.   Major schemes like National mission for sustainable agriculture, Rain fed area development scheme, Submission on Agro forestry, Soil Health Management and Param paragat Krishi Vikas Yojana are few of the government schemes which are enabling for Sustainable agriculture development in India.

Keywords: Green Revolution, Organic Agriculture, Soil Health, Rainfed Area, Agro Forestry.

Introduction

Agriculture is a basic necessity for survival of human being. Indians have been following traditional way of agriculture since the dawn of civilization. Indian government has initiated green revolution in order to meet the food shortage faced by India during 1960s. The green Revolution adopted modern industrial system in agriculture leading to use of high yielding variety seeds, chemical pesticides and fertilizers along with that advanced mechanized farm tools. Although Green Revolution has given a short term success in production of food crops, increased the income of the large farmers, increase in the demand for rural labours leading to employment opportunities in rural areas and reduce the import of the food grains.

In the long run green revolution has created shortage of other food grains such as core cereals pulses and oil seeds, it has led to wide regional disparities. Excessive usage of chemicals associated with the green revolution has created high risk causing harm to environment and creating soil pollution leading to the depletion of agriculture and sustainability of agriculture. Post Green Revolution Indian agriculture is facing new challenges. The Government of India at Central and various state governments are engaged in addressing the challenges posed due to various factors. The second generation problems of farming community such as sustainability, adoption of new technologies in farming and nutrition related challenges are addressed by reframing various policies and programs at various departmental levels of the government.

The present research paper wants to analyze the schemes which are framed towards development of sustainable agriculture in India. The present research paper wants to analyze the schemes which are framed towards development of sustainable agriculture in India.The Government of India has brought many programs, policies and plans in order to meet the sustainable development goals envisioned by the United Nations, especially goals related to the goal of ending hunger achieving food security and improving nutrition and to promote sustainable agriculture. Let us analyze the programs, plans and policies implemented by the government of India in realizing sustainable agriculture development.

Objective of the paper

The present research paper wants to analyze the schemes which are framed towards envisioning development of sustainable agriculture in India.

Research Methodology

This paper is presented by reviewing various secondary sources like books, journals, Census reports and various other webliographical sources.

Government Efforts in Promoting Sustainable Agriculture in India

The Present research paper wants to know the schemes, plans and policies which are striving towards creating sustainable agriculture environment in India. Agriculture is placed in state list under the ambit of Indian Constitution. The policies plans and programs carved out by the central government should be implemented by the states to suit their needs. Keeping these things in mind now let us analyze the schemes which are implemented for promoting sustainable agriculture in India.

National mission for sustainable agriculture

National Action Plan on climate change is a flagship program of Government of India implemented in the year 2008 in order to control and address the negative effects of climate change. The Action Plan consists of eight major missions and National mission for sustainable agriculture is one of the prominent among the eight missions which is focusing on the sustainable agricultural growth. During XII Five Year Plan National mission for sustainable agriculture focuses on 10 key principles of Indian agriculture. They are water use efficiency, pest management, improved crop seeds, credit support, agricultural insurance, livestock and fish cultures, markets, livelihood diversification and access to information. The major focus of national mission on sustainable agriculture is to use resources judiciously through community based approach. The main objective of national mission on sustainable agriculture is to make agriculture more sustainable, productive, climate resilient followed by healthy practices to conserve natural resources and keep soil fertility and optimum utilization of water through comprehensive soil Health and efficient water Management. The ultimate aim of the national mission on Sustainable agriculture is summarized in the Annual report of Ministry of Agriculture and Farmers Welfare 2019, that, "making agriculture more productive, sustainable, remunerative and climate resilient. This can be achieved by promoting location specific integrated/composite farming systems; soil and moisture conservation measures; comprehensive soil health management; efficient water management practices and mainstreaming rain fed technologies"   MoAFW 2019: 101. The National mission for sustainable agriculture has four major components to address and to improve the sustainable agricultural dimensions in India through following Government of India sponsored schemes. They are as follows:

Rainfed area development

Rainfed area development scheme is an intensive scheme to make rainfed agriculture more income oriented, effective, productive sustainable and climate resilient by promoting mixed cropping practices such as combining agriculture with horticulture, livestock rearing, fisheries etc. The major of the schemes to improve farmer's income and to curb the impact of floods and drought on crops. Table 1.1 gives us the information about the targets given to the rainfed area development scheme and the actual achievements year wise. From the table we can interpret that the scheme is not up to the mark and failing the actual targets set by rainfed area development scheme.

Submission on Agroforestry (SMAF)

This mission is launched in the year 2016 with the main intention of increasing tree plantation along with crop production. The scheme focuses on promoting Agro forestry by growing trees and crops responding to local environmental conditions.

TABLE 1.1: Target and achievement under rainfed area development scheme

Year	Coverage under Integrated Farming System (in Hectare) Target	Coverage under Integrated Farming System (in Hectare) Achieved	Expenditure in Crores Target	Expenditure in Crores Achieved
2015-16	42,380.6	35543.1	152.2	123.5
2016-17	55,833.5	40961.5	200.8	147
2017-18	72,518.3	50075.8	270.1	180.2
2018-19	99,346.5	69988.8	281.1	183.4
2019-20	50,114.6	45270.5	191.1	66.6

Source: NMSA

This mission is contributing in enhancing soil health and increasing farmer's income another agenda of the scheme is to increase trees ratios and create a pool of trees which help in mitigating carbon dioxide and creating natural carbon sink which is one of the prominent conditions in Paris agreement. Table Number 1.2 shows the effort made under this project since its initiation. From the table we can analyze that in the year 2016-17 it started with coverage of 373 hectare area followed by rampant increase in the year 2017 18 and 2018-19 cross coverage of more than 2000 hectare area and drastically reduced in the year 2019-20 covering only 774 hectare area leading to decline in the year 2019-20 compared to two previous years.

TABLE 1.2: Annual area covered and tree planted under SMAF

Year	Area covered (hectare)	Trees Planted
2016-17	373	4,63,159
2017-18	2273	34,73,699
2018-19	2819	19,37,075
2019-20	774	2,60,056

Soil health management

The major intention of this scheme is to promote regional as well as crop specific sustainable soil Health Management along with organic farming practices, Residue management, macro micro nutrient management, judicious use of land and application of fertilizers which helps in decreasing soil erosion. The schemes gives extensive tasks to state governments, soil and land use Survey of India, Central fertilizer quality control and training institute, National centre of organic farming all these major institutions are involved in creating sustainable soil Health Management throughout India.

Under this ambitious mission soil health card scheme is established and distributed to the farmers and check their nutrients status of their land. Hence the soil health card scheme wants to improve nutritional status of the soil along with suggestions to improve organic content of the soil over the time. A wide range of distribution of soil health cards is issued under this scheme. In the year 2015-17 about 107.3 million farmers got soil health card.

Param paragat krishi vikas yojana (PKVY)

Param paragat Krishi Vikas Yojana is a subsidiary of Mission soil Health Management established under National Mission on Sustainable agriculture. The major aim of the param paragat Krishi Vikas Yojana is to practice traditional agriculture along with the modern scientific lines, so that the soil fertility couldn't be affected much. The scheme also aims to enhance production of quality foods with the zero usage of chemicals and pesticides.

The scheme wants to enhance the income of the farmers by following cluster method in marketing. The funding pattern of this scheme is 90:10 i.e. 90 percentage Centre funded and 10 percent state funding. The thrust area of this scheme is to promote organic farming among the youth farmers of rural area by adopting eco friendly and rural Technologies.Scheme has benefited around 8.89 lakh farmers and due to this scheme the net cultivated area of organic farming is increasing year by year.

Conclusion

The above listed schemes are playing crucial role in creating a sustainable agriculture environment in India. There are many subsidiary schemes which are run by collaborating various agencies Central Government, State Government, ministry of agriculture department of agriculture and farmers welfare,department of horticulture, department of land use and survey and various other University's of Agricultural Sciences also involved in creating the sustainable agriculture development environment in India.

References

- Ministry of Agriculture and Farmers Welfare, (2016). Sub-Mission on Agroforestry: Operation Guidelines. https://nmsa.dac.gov.in/pdfdoc/Agroforestory_Guidelines_English.pdf.
- Ministry of Agriculture and Farmers Welfare, (2017). National Mission for Sustainable Agriculture: Operational Guidelines. Government of India.
- National Mission for Sustainable Agriculture (NMSA), Department of Agriculture and cooperation, Ministry of Agriculture New Delhi August 2010.
- PKVY, (2017). Param paragat Krishi Vikas Yojana: Manual for District-Level Functionaries.
- https://darpg.gov.in/sites/default/files/Paramparagat%20Krishi%20Vikas%20Yojana.pdf.
- Soil Health Card (n.d.). "Scheme Progress".
- https://soilhealth.dac.gov.in/publicreports/dashboardtargetreport.

Chapter 15

ISBN: 978-81-948672-7-2

An Analysis of Sustainable Agricultural Development in India

DR. RAGHAVENDRA B. N.
Principal
Harsha Institute of Management Studies,
Email: bnraghavendra2005@gmail.com

Abstract: The new agenda in Indian agriculture should have a goal that explicitly focuses on improving agricultural systems and addresses rural development in an integrated manner. While Indian agriculture has crossed the threshold of traditional farming to modern agri-business, the objective of ensuring equity and sustainability becomes all the more important. Agriculture must change to meet the rising demand, to contribute more effectively to the reduction of poverty and malnutrition, and to become ecologically more sustainable. The challenge is daunting but feasible. This paper analyses the issues related to sustainable agriculture development. There is hope for positive changes as more and more farmers are becoming open-minded and government is coming up with new initiatives. The study suggests that problems largely institutional, structural and administrative – need to be overcome for development in general and sustainable agricultural development in particular.

Keywords: Sustainable Development, Farming, Agriculture, Malnutrition.

Introduction

Agriculture in India is facing several challenges which together manifest into sustainability. It makes direct use of natural resources. It is separated from the tertiary sector (which creates completed goods) and the secondary sector (which produces manufactured and other processed items). In general, this industry is more prevalent in developing countries and less so in wealthy countries. Agriculture employed the vast majority of the human population until the industrial revolution.

In pre-industrial agriculture, farmers raised the majority of their crop for their own consumption rather than as a cash crop for exchange. The Indian economy's most important sector is agriculture. It has seen tremendous transformation in the previous two decades, with policies of globalisation and liberalisation opening up new chances for agricultural modernization. Because of the investments made in the in general, this industry is more prevalent in developing countries and less so in wealthy countries.

Agriculture employed the vast majority of the human population until the industrial revolution. In pre-industrial agriculture, farmers raised the majority of their crop for their own consumption rather than as a cash crop for exchange. The Indian economy's most important sector is agriculture. It has seen tremendous transformation in the previous two decades, with policies of globalisation and liberalisation opening up new chances for agricultural modernization. Not only has investment in the sector resulted in commercialization and diversification, but it has also resulted in numerous technological and institutional advancements. This industry is particularly prevalent in poorer countries. "Agriculture has always been the backbone of the Indian economy," according to the report, and despite fears of industrialization, agriculture has remained a source of national pride for the past six decades. It employs over 60% of the total workforce in the country. The success of the country's economy is dependent on agricultural expansion. Villages 3244 are home to around 67 percent of India's people who work for a living.

Their principal "Agriculture is the largest and most visible sector of our economy," he says, noting that it employs roughly 70% of India's population. Traditionally, agricultural products have been the principal source of raw materials. This illustrates agriculture's perceived importance and domination in the "Indian economy." It also contributes significantly to GDP. Agriculture productivity growth, in both the agricultural and non-agricultural sectors, is a major driver of economic growth and poverty reduction.

Unfortunately, food and agriculture are once again at the forefront as a result of poverty connected to price instability, and growing food costs are on the rise. In recent years, a global race for agricultural land, large private investments in agriculture enticed by attractive markets, obstructive import tariff policies in food staples, the renewal of large-scale agricultural subsidies in the name of food security, and new beneficiary responsibilities to increase agricultural expenditure after 20 decades of neglect have all been developed to address the issue of food security. The main issue is figuring out how to use agriculture to help support a structural shift in the economy, As a result, we began by examining agriculture's role in the development process and its connections to other industries. Agriculture growth has the ability to relieve poverty in emerging countries dramatically.

Objectives of the Study:
The following were the primary objectives of this research
1. To analyse the contribution of agricultural sector to the Indian economy's GDP.
2. To assess the Government Programs towards agricultural sustainable development.
3. To examine the India's policy ecosystem for sustainable agriculture.

Methodology of the study

The data for this study was gathered completely from secondary sources. Information obtained by someone other than the original user is referred to as secondary data. Censuses, government data, organizational records, and data obtained for other research purposes are all common sources of secondary data in social science. In the case of quantitative data, it is possible to create larger and higher-quality databases than a single researcher could obtain on their own.

Contribution of Agricultural Sectorto India's GDP

India's largest industry is the services sector. In 2020-21, the services sector's Gross Value Added (GVA) is expected to be 96.54 lakh crore INR at current prices. The services industry contributes for 53.89 percent of India's overall GVA, which is worth 179.15 lakh crore rupees.

Industry provides 25.92 percent of GDP, with a GVA of Rs.46.44 lakh crore. Agriculture and related industries account for 20.19 percent of the total. Agriculture & allied, Industry, and Services make up 16.38 percent, 29.34 percent, and 54.27 percent of the economy, respectively, at 2011-12 prices. Primary (agricultural, forestry, fishing, and mining & quarrying) and secondary (manufacturing, electricity, gas, water supply & other utility services, and construction) sectors are anticipated to account for 21.82 percent, 24.29 percent, and 53.89 percent of GDP, respectively.

At current prices in 1950-51, the proportions of Agriculture & allied, Industry, and Services were 51.81 percent, 14.16 percent, and 33.25 percent, respectively, according to prior methods. Agriculture and allied sector's share of GDP fell to 18.20 percent in 2013-14. The Services sector's share has increased to 57.03 percent. The industry sector's share has also risen to 24.77 percent. According to the CIA Facebook, India's GDP composition by sector in 2017 was as follows: Agriculture (15.4%), Industry (23%), and Services (23%). India is the world's second largest producer of agricultural products, with $375.61 billion in production.

India produces 7.39 percent of the world's total agricultural output. India lags well behind China, which has a $991 billion GDP in agriculture. Industry's GDP is $560.97 billion, and it ranks 6th in the world. India is ranked eighth in the world in the services industry, with a GDP of $1500 billion. The agricultural industry contributes significantly more to the Indian economy than the global average (6.4 percent). The contribution of industry and services is lower than the global average of 30% for industry and 63 percent for services.

Gross Value Added (GVA)

	Sector	GVA in 2020-21 (Rupees in Crore)			
		Constant Prices	Share (%)	Current prices	Share (%)
1	**Primary Sector**	**2,334,723**	**18.75%**	**3,908,643**	**21.82%**
1.1	Agriculture, Forestry &Fishing	2,040,079	16.38%	3,616,523	20.19%
1.2	Mining &Quarrying	294,644	2.37%	292,120	1.63%
2	**Secondary Sector**	**3,359,718**	**26.98%**	**4,352,265**	**24.29%**
2.1	Manufacturing	2,107,068	16.92%	2,585,740	14.43%
2.2	Electricity, gas, water supply &other utility services	306,254	2.46%	484,477	2.70%
2.3	Construction	946,396	7.60%	1,282,048	7.16%
3	**Tertiary Sector**	**6,758,989**	**54.27%**	**9,654,259**	**53.89%**
3.1	Trade, Hotels, Transport, Communication and Services related to broadcasting	2,208,388	17.73%	2,941,477	16.42%
3.2	Financial real estate &Profservs	2,872,815	23.07%	3,950,786	22.05%
3.3	Public Administration, defence and other services	1,677,786	13.47%	2,761,996	15.42%
	GVA at basic prices	**12,453,430**	**100%**	**17,915,167**	**100%**

Factor influencing Sustainable Agriculture in India

Income: There is insufficient evidence on the influence of SAPSs on farmer incomes, both in terms of geographical coverage and the number of long-term assessments. Despite this significant constraint, the literature suggests that a few SAPSs have the ability to increase revenue by lowering production costs (CA, natural farming), diversifying agricultural production (IFS, intercropping), and charging premium prices (organic produce).

Yields: We identify some emergent patterns for yields under a few SAPSs, despite the conceptual limits of accurately estimating agricultural output. Organic agricultural yields are lower than conventional farming, at least in the short term (2-3 years). Some studies demonstrate that after this period, some crops can provide equal or even better yields, especially if the soil form and structure are improved. For most crops, short-term trials of natural farming show no statistically significant changes in yields. The effects of SRI on yield have been thoroughly documented, with statistically significant increases in many paddy kinds. Vermicomposting, agroforestry, and crop diversification are examples of resource-saving strategies that have improved yields. However, because there are few studies on the long-term effects of SAPSs on yields, it's difficult to extrapolate conclusions.

Water use: Several research have been published that show how different SAPSs affect water consumption efficiency. Water conservation has been aided by SRI, CA, precision farming, rainwater collection, contour farming, cover crops, mulching, crop rotation, and agroforestry. Smallholder farmers like rainwater collecting and SRI because they are simple to implement. In drought-prone Andhra Pradesh, pre-monsoon dry sowing in natural farming is seen as a breakthrough that warrants further investigation.

Bio-diversity: Agroforestry, IFS, permaculture, natural farming, organic farming, conservation agriculture, and crop diversification tactics (rotation, intercropping, mixed) all tend to improve the spatial, vertical, and temporal diversity of species on a farm (and landscape) level. While study publications discuss the influence on biodiversity, there are no studies that provide solid empirical data.

Health: We only have anecdotal evidence that many SAPSs have favourable health effects, primarily due to nutritional diversification and less exposure to dangerous substances like pesticides. There are no empirical studies comparing SAPSs to traditional agriculture in terms of health consequences.

Gender: In Indian agriculture, women make up more than 70% of the workforce. However, there are few studies that focus on the gender consequences of SAPSs. Women's responsibilities are defined by a few techniques such as vermicomposting, organic farming, IFS, and rainwater gathering, although evidence on their influence is lacking. To fully comprehend the influence of various SAPSs on women's workloads, income, empowerment, and employment, more research is required.

India's policy ecosystem for Sustainable Agriculture

India has had a National Mission for Sustainable Agriculture (NMSA) in place to promote sustainable agriculture since 2014-15. It is made up of a number of programmes that focus on agroforestry, rainfed areas, water and soil health management, climate impacts, and adaptation. Aside from NMSA, the Pradhan Mantri Krishi Sinchai Yojana encourages precision farming techniques like micro-irrigation, while the Integrated Watershed Management Program encourages rainwater gathering. NMSA, on the other hand, receives only 0.8 percent of the Ministry of Agriculture and Farmers Welfare's (MoAFW) budget. Aside from the MoAFW's budget of 142,000 crores ($20 billion), the Central government spends roughly 71,309 crores ($10 billion) on fertiliser subsidies each year.

As a result, while the Indian government recognizes the necessity of supporting sustainable agriculture, it continues to place a high priority on it. Eight of the 30 SAPS practises receive some budgetary support from the federal government through various programmes.

Organic farming, integrated agricultural systems, rainwater harvesting, contour farming (terraces), vermicomposting, mulching, precision farming, and IPM are examples of these practises. Organic farming has gotten the most policy attention of them, with some Indian states enacting exclusive organic farming regulations.

These CSOs provide a variety of services to promote SAPSs, including farmer training, capacity building, and awareness raising, input preparation and seed management assistance, and field demonstration activities. A small number of people are also involved in technology transfer.

The National Mission for Sustainable Agriculture receives only 0.8 percent of the Ministry of Agriculture and Farmers Welfare's budget, indicating a considerable scope.

Government Scheme for Sustainable Development:
"Pradhan Mantri Kisan Samman Nidhi (PM-KISAN) Scheme": This project claims to pay all subsistence farmers Rs 6,000 per year in three instalments via Direct Bank Transfer (small and marginal farmers with lands of up to 2 hectares). According to sources, it is intended to help 14.5 crore farmers across India.

"Pradhan MantriKisan Pension Yojana" means "Pradhan Mantri Kisan Pension Yojana: The "Modi 2.0 Cabinet" agreed to a plan to compensate small and marginal farmers with a monthly fixed pension of at least Rs 3,000, costing the exchequer Rs 10,774.5 crore per year, in order to ease hardship in the agriculture industry.

"Pradhan Mantri Jan DhanYojana: is a "National Financial Inclusion Mission" that takes a holistic strategy to achieve "complete financial inclusion" and providing banking services to all families.

"National Mission for Sustainable Agriculture (NMSA)": To improve agricultural output, it focuses on intensive agriculture, water consumption efficiency, soil health management, and resource conservation harmony, particularly in rainfed areas.

"Pradhan Mantri Krishi Sinchayee Yojana (PMKSY)": Water conservation and management are important priorities for the Indian government. A plan has been established to achieve this goal, with the purpose of increasing irrigation capacity. The "NDA government" founded it in 2015 as a government programme to support organic farming in the country.

"Pradhan Mantri Fasal BimaYojana (PMFBY)": It is a "government-sponsored crop insurance programme" that brings several stakeholders together on a single platform. This programme aimed to provide farmers and cattle breeders with a way to protect themselves against the loss of their animals due to death, as well as to demonstrate to the general public the benefits of livestock insurance, with the ultimate goal of improving the quality of livestock and their products.

Conclusion

This research looked into the intricacies of agricultural development, which is critical for low-income developing countries' economic growth, rural development, and poverty reduction. Agricultural productivity improvements are a crucial driver of economic growth and poverty reduction both inside and beyond the agricultural industry. Appropriate infrastructure, well-functioning domestic markets, competent institutions, and access to appropriate technologies are all required in order to boost productivity. Agriculture is India's most important source of revenue. Agriculture accounts for around 20.19 percent of the country's gross domestic product. Despite the fact that it remains the largest donor, the agriculture sector's contribution has dropped in recent years. Agriculture is extremely important in Indian economics, not only because it contributes to "GDP," but also because it employs a large section of the population. On the other hand, India's agriculture sector appears to be well on its way to realising its full potential. The green revolution increased food grain output while simultaneously offering agricultural technology improvements. This is reflected in India's improved net trade balance.

Since 1990, India has been a net exporter of agri-food goods, where it previously relied on imports to feed its people. Since 1990, India has been a net exporter of agri-food products, when it previously relied on imports to feed its people. The rural non-farm industry has considerable potential, notwithstanding its connections to small cities and rural areas. Rural development and community-driven development can help to enable this strategy.

The government is required to concentrate and contribute for the growth of agricultural sustainability in India. The private sector will be the key source of investment capital as well as a provider of services. Non-governmental organisations (NGOs) and civil society organisations (which benefit from local and international private knowledge when implementing programmes) will also play a key role.

In order to improve agricultural and rural development, the Indian government prioritises poverty reduction through higher agricultural output. Policymakers will need to take major steps to move away from the current "subsidy-based system," which is no longer economically feasible, in order to build a "strong foundation for a highly productive, internationally competitive, and diverse agricultural business."

References:

- G. Pal, K. S. Dipak, and Indusnettechnologies, "Home: Agriculture Department, Government Of Uttarakhand, India." [Online]. Available: http://www.agriculture.uk.gov.in/.[Accessed: 10-Jul-2020]
- T. Partap, "Hill agriculture: challenges and opportunities." Indian Journal of Agricultural Economics 66, no.1 902-2016-67891, 2011.
- U. Tuteja, "Agriculture Profile of Uttarakhand", Agricultural Economics Research Centre, University of Delhi, 2013.
- M. V. Ramesh, R. Mohan, and S. Menon, "Live-in-Labs: Rapid translational research and implementation-based program for rural development in India," in GHTC 2016 - IEEE Global Humanitarian Technology Conference: Technology for the Benefit of Humanity, Conference Proceedings, 2016.

- Amrita SeRVe - Self Reliant Village : https://amritaserve.org/Amrita SeRve, 'The first years report' [Online] https://amritaserve.org/wpcontent/ uploads/2019/05/2019AmritaSeRV eTheF irst5Y ears.pdf
- T. Brown and J. Wyatt, "Design thinking for social innovation." Development Outreach 12, no. 1, pp.29-43, 2010.
- B.J. Ranger and A. Mantzavinou. "Design thinking in development engineering education: A case study on creating prosthetic and assistive technologies for the developing world." Development Engineering 3,pp.166-174, 2018.
- K. Hemasagar, S. Eleshwaram, R. Mohan, S. Ariprasath, K. Nandanan, S.G.D. Sharma, and B. Siddharth, "Water Management Through Integrated Technologies, a Sustainable Approach for Village Pandori,India." In 2019 IEEE R10 Humanitarian Technology Conference (R10-HTC)(47129), pp. 180-185, IEEE, 2019.

Chapter 16   **ISBN: 978-81-948672-7-2**

Rural Development and Sustainable Agriculture Development

M.N.MADHURA
Assistant Professor,
Department of Economics
Basudev Somani College, Vishwamanava Double Road,
Kuvempunagar, Mysore -570023
Email: madhruvidwath13@gmail.com

Abstract: Sustainable agriculture and rural development [SARD] play a crucial role in maintaining natural resources, protecting the ecological environment, enhancing global food security, eradicating poverty and even promoting rural revitalization, rural empowerment and environment management. In order to solve the problems of rural development, agriculture extension systems needs to encourage the active participation of farmers in planning, implementing and monitoring agricultural extension programs.

Keywords: Decentralization, Pluralism, Sustainable Agricultural development.

Introduction

The food and agricultural organization define sustainable agriculture development is "the management and conservation of the natural resources, base, and the orientation of technological and institutional change in such a manner as to ensure the attainment and continued satisfaction of human needs for present and future generations. Such development conserves land, water, plant and animal genetic resources are environmentally none degrading, technically appropriate, economically, viable and socially acceptable". The challenges of increasing food production are daunting. Despite great agricultural advances, millions go hungry or live under threat of famine. Food production will have to double between 1995 and the year 2025 if the expected population of up to 8500 million is to be fed adequately. Farmer's involvement is the key to sustainable agriculture. Given the right incentives and Government support, farm families can and are making significant progress towards managing their land and water sustainably.

Agriculture plays an important role in any economy. It is directly and indirectly connected with the economic activity, growth and development of other sectors in economy and on the whole welfare and development of an economy. Agriculture is one of the productions, both food for the rural and urban population and of cash crops for the export market to earn foriegn currency.

India is an agrarian economy and agriculture sector has still a lot of bearings on the overall growth and development of the country normally and rural development particularly. The economic division of agriculture to India's gross domestic product [GDP] is declining with the countries broad based economic growth.

Agriculture is demographically the broadest economic segment and plays a major portion in the overall socio-economic fabric of India. Agriculture with its allied sectors is unquestionably the main livelihood provider in India.

So, in the vast rural areas 69% of India's population lives in rural regions and 3/4 of the people making up these rural populations depend on agriculture and allied activities for their livelihoods.

Objectives of rural development:

1. To improve the living standards by providing food, shelter, clothing, employment and education.
2. To increase productivity in rural areas and reduce poverty.
3. To involve people in planning any development through their participation in decision making and through centralisation of administration.
4. The achievement of the millennium development goals is at the centre of sustainable development. Sustainable rural development is vital to the economic, social and environmental viability of nations.

Components of sustainable agriculture:

- Soil management, crop management, water management, disease or pest management and waste management are the main components of sustainable agriculture.

- Future sustainability can be determined by the most limiting elements and these elements may change over time including population density, water, fossil fuel energy, nitrogen, carbon dioxide, salinity, economy and agricultural land quantities.

- Good soil management practices are needed to achieve the sustainability principle, including spreading fertilizers, implementing composts, planting cover crops and decreasing tillage. Best crop management practices include pinching, chopping, hilling and suckering. The best practices for water management include drip irrigation, rotational grazing, crop covering, dry farming water capture and storage irrigation scheduling.

Why is sustainable agriculture important to us?

Important because it means human food, fuel and fibre needs while maintaining habitat conservation and biodiversity protection. Sustainable agriculture is essential in our lives and plays an important role in the future because it is related directly and indirectly to our food security and connected with Economic, social and environmental factor.

- To Promote Social innovations for sustainable rural development with inclusive transformation
- To integrate rural development strategies into poverty reduction strategies
- To promote social capital and resilience in rural communities
- To Promote and ensure equitable access to land, water and financial resources
- Active participation of vulnerable groups and to safeguard their livelihood
- To encourage rural communities, participation in decision making, promote rural communities, in particular for youth, young girls, women and indigenous people.

- Alternative and renewable reliable sources of energy for sustainable development in rural areas.
- Support of both agriculture and non-agriculture services towards sustainable development
- To assess Equal opportunities for women and men in all aspects of rural development.
- Promote sustainable natural resources use and management including ecosystem conservation through community-based Institution based organisations.
- To make rural areas and human settlements inclusive, safe, resilient and sustainable.
- Policies and strategies provide the required coherence by the government.
- To draw pathway towards Sustainable Development Goals 2030
- Access to knowledge, information and education, political empowerment of people, equity, sustainability, attitudes and values that foster responsibility, solidarity and tolerance.
- To promote complementary economic activities that could further stimulate rural entrepreneurship while decreasing rural community dependency on one main economic sector.

Sustainable Development Goals (SDGs):

The Sustainable Development goals (SDGs) are successors to the 'Millennium Development Goals MDGs'. The MDGs were adopted in 2000 by governments to make global progress on poverty, education, health, hunger and the environment. The MDGs expired at the end of 2015. During 25-27 September 2015, the member states of the United Nations converged in New York for the United Nations (UN) Summit on Sustainable Development and adopted the new global goals for sustainable development.

The world leaders pledged their commitment to the new '2030 Agenda for Sustainable Development', encompassing 17 universal and transformative SDGs. The United Nations General Assembly has taken the resolution, for adopting 17 Sustainable Development Goals (SDGs), with 169 targets and 304 indicators, on 25th September 2015 under the official agenda Transforming our world: the 2030 Agenda for Sustainable Development".

The SDGs are a universal set of goals, targets and indicators that all UN member states are expected to use to frame their development agendas, socio-economic policies, and actions towards low carbon pathways for the next 15 years, in order to achieve a sustainable world where 'no one is left behind' without compromising sustainability of the planet.

These new global goals are much broader and comprehensive than the outgoing MDGs, as they attempt to address all three dimensions of sustainable development- economic, social and environmental. The list of 17 SDGs adopted by the United Nations is:

- Poverty - End poverty in all its forms everywhere.
- Food - End hunger, achieve food security and improved nutrition and promote sustainable agriculture.
- Health - Ensure healthy lives and promote well-being for all at all ages.
- Education - Ensure inclusive and equitable quality education and promote lifelong learning opportunities for all.
- Women - Achieve gender equality and empower all women and girls.
- Water - Ensure availability and sustainable management of water and sanitation for all.
- Energy - Ensure access to affordable, reliable, sustainable and modern energy for all.
- Economy - Promote sustained, inclusive and sustainable economic growth, full and productive employment and decent work for all.
- Infrastructure - Build resilient infrastructure, promote inclusive and sustainable industrialization and foster innovation.
- Inequality - Reduce inequality within and among countries.
- Habitation - Make cities and human settlements inclusive, safe, resilient and sustainable.
- Consumption - Ensure sustainable consumption and production patterns.
- Climate - Take urgent action to combat climate change and its impacts.
- Marino systems - Conserve and sustainably use the oceans, seas and marine resources for sustainable development.

- Ecosystems - Protect, restore and promote sustainable use of terrestrial ecosystems, sustainably manage forests, combat desertification, and halt and reverse land degradation and halt biodiversity loss.
- Institutions - Promote peaceful and inclusive societies for sustainable development, provide access to justice for all and build effective, accountable and inclusive institutions at all levels.
- Sustainability - Strengthen the means of implementation and revitalize the global partnership for sustainable development.

Conclusion:

It draws attention to the societal and environmental threats which rural communities around the world are facing. Agenda 2030 and SDGs aim to eradicate extreme poverty, famine, open defecation, and other critical issues in developing countries associated with lack of public utilities, mainly in rural areas, and to reduce the huge gaps between countries and regions. To achieve all range of SGDs across the globe, proper attention must be paid to rural development perspectives such as quality of life improvement, sustainable agriculture, rural resilience, and circular economy and reduced inequalities.

Sustainable rural development involves a holistic approach where daily basic needs of rural populations must be covered by reliable public utilities combined with technical, socioeconomic, and environmental conditions to support regional economies and urban-rural linkages. Rural communities must develop several nonfarming activities coupled with agricultural systems (adapted to local geographical conditions) to become more resilient to economic shocks or environmental disturbances in the context of climate change. Rural areas should receive the same attention and opportunities from decision-makers, academics, and professionals regarding sustainable development policies and investments in infrastructure projects. Agenda 2030 could be achieved if sustainable rural development policies will be implemented in each country next to urban areas. Summing Up It draws attention to the societal and environmental threats which rural communities around the world are facing. Agenda 2030 and SDGs aim to eradicate extreme poverty, famine, open defecation, and other critical issues in developing countries associated with lack of public utilities, mainly in rural areas, and to reduce the huge gaps between countries and regions. To achieve all range of SGDs across the globe, proper attention must be paid to rural development perspectives such as quality of life improvement, sustainable agriculture, rural resilience, and circular economy and reduced inequalities. Sustainable rural development involves a holistic approach where daily basic needs of rural populations must be covered by reliable public utilities combined with technical, socioeconomic, and environmental conditions to support regional economies and urban-rural linkages.

Rural communities must develop several non farming activities coupled with agricultural systems (adapted to local geographical conditions) to become more resilient to economic shocks or environmental disturbances in the context of cli1mate change. Rural areas should receive the same attention and opportunities from decision-makers, academics, and professionals regarding sustainable development policies and investments in infrastructure projects. Agenda 2030 could be achieved if sustainable rural development policies will be implemented in each country next to urban areas.

References:

1. Justice Mensah., Sandra Ricart Casadevall (Reviewing editor) (2019). Sustainable development: Meaning, history, principles, pillars, and implications for human action: Literature review, Cogent Social Sciences, 5:1.

2. C. Dankers, and P. Liu, FAO: Environmental and social standards, certification and labelling for cash crops, Commodities and Trade Technical Paper No. 2. Rome, 2003.

3. Moseley, Malcolm J. (2003). Rural development : principles and practice (1. publ. ed.). London [u.a.]: SAGE. p. 7. ISBN 978-0-7619-4766-0.

4. Van Assche, Kristof. & Hornidge, Anna-Katharina. (2015) Rural development. Knowledge & expertise in governance. Wageningen Academic Publishers, Wageninge.

5. https://www.intechopen.com ›

6. Infrastructure and Sustainable Development Issues and Challenges : Editor in Chief –Dr.M.Mahadevaiah,Published by: Basudev Somani College,kuvempunagar,mysore-23.Published on January 19,2018.